04-2017

THE GROUND
BENEATH US

ALSO BY PAUL BOGARD

The End of Night

THE GROUND BENEATH US

BENEATH US

From the Oldest Cities to the Last Wilderness,
What Dirt Tells Us About Who We Are

PAUL BOGARD

Little, Brown and Company

New York • Boston • London

Little, Brown and Company
Hachette Book Group
1290 Avenue of the Americas, New York, NY 10104
littlebrown.com

First Edition: March 2017

Little, Brown and Company is a division of Hachette Book Group, Inc. The Little, Brown name and logo are trademarks of Hachette Book Group, Inc.

The publisher is not responsible for websites (or their content) that are not owned by the publisher.

The Hachette Speakers Bureau provides a wide range of authors for speaking events. To find out more, go to hachettespeakersbureau.com or call (866) 376-6591.

ISBN 978-0-316-34226-1
Library of Congress Control Number: 2016958940

10 9 8 7 6 5 4 3 2 1

LSC-C

Printed in the United States of America

For Caroline

Talk of mysteries! — *Think of our life in nature,* — *daily to be shown matter, to come in contact with it,* — *rocks, trees, wind on our cheeks! The* solid *earth! the* actual *world! the* common sense! Contact! Contact! Who *are we?* where *are we?*

— HENRY DAVID THOREAU, *THE MAINE WOODS* (1864)

Contents

THE GROUND BENEATH US

Introduction

For a transitory enchanted moment man must have held his breath in the presence of this continent, compelled into an aesthetic contemplation he neither understood nor desired, face to face for the last time in history with something commensurate to his capacity for wonder.

    — F. Scott Fitzgerald, *The Great Gatsby* (1925)

From the observatory deck on the 102nd floor of One World Trade Center, the island of Manhattan spreads north beneath a hazy gray-gold sky — the Hudson River on the left, the East on the right — a putty-white and beige-brown blanket of stone, steel, and concrete covering every inch of its once green ground. A belt of Midtown towers blocks the view of Central Park's verdant rectangle, but I know it's there on the other side. Just as I know what lies beneath the city's weight. "I became aware of the old island here that flowered once for Dutch sailors' eyes," Fitzgerald wrote, "a fresh, green breast of the new world."

From near the top of the tallest building in the Western Hemisphere, as far from ground as it's possible to be while still connected, it's difficult to see more than a few scraps of that old new world. But my plan is to walk up the island from Ground Zero to the park. Here, in my country's largest city, its most urban area, I will begin to look for the ground beneath us—what's gone missing, what remains, what may come to be.

For now, what is under my feet here on the observation deck is turning my legs rubbery. Ninety million pounds of structural steel and more than two hundred thousand cubic yards of concrete—enough for a sidewalk stretching from here to Chicago—hold me up and anchor this tower down, but still I swear I can feel it sway. Long before the tower rose, the below-ground level had to be prepared. Workers cleared debris from 9/11's wreckage and dug into the bedrock that would support the new building—two hundred feet below street level. Digging that deep yielded immediate discoveries, including shoes, wallets, and even human remains. Deeper still, workers struck timber, the oak ribs of an eighteenth-century boat from an era when the Hudson flowed through the site. Is it knowledge of the depth and weight of foundational construction that gives most people the confidence needed to press against the thirty-foot-tall floor-to-ceiling window glass, expressing amazement in languages from across the globe as they peer 1,268 feet down? While I do appreciate the view, I cannot wait to get back to solid ground.

Less expansive than "earth," less ambiguous than "the land," for me "ground" means where we "have trod, have trod," as Gerard Manley Hopkins wrote in his famous poem, and continue to tread, what we see when we look down, the planet as we experience it in our day-to-day lives.

And it is a wonderland. We walk on ground that teems with life—an incredible one-third of all living organisms—a trove of biodiversity still only just starting to be explored. Said Leonardo da Vinci some five hundred years ago, "We know more about the movement of celestial bodies than about the soil underfoot." Incredibly, that remains true. But it's also true that this is changing. Since the new century's start, our knowledge of soil has bloomed—we have learned more in the past decade than in all previous years combined. We now know, for example, that just a teaspoon of healthy soil holds millions of species, and far more microorganisms than there are people on Earth. This incredible living ground gives rise to every plant, animal, and human—some 97 percent of the food we eat comes from the ground—and to boundless beauty. From the tallest redwoods and evergreens to the tiniest blue wildflowers, the ground holds the wild world in place. The paths we walk repeatedly, the pressed prints of elephant, wolf, and lion, the bodies of those who fought and fell, the memories of everything gone before, the ground holds this all.

Unfortunately, studies reveal that most of us in the industrialized world spend 90 to 95 percent of our time indoors. And when we do walk outside, we see beneath our shoe-clad feet an unnatural surface, likely some version of asphalt or concrete. In fact, we have some sixty-one thousand square miles of paved ground in the United States, an amount that together would be the twenty-fourth largest state by surface area, larger than any state east of the Mississippi. We now have more square miles of pavement in the lower forty-eight than we have square miles of wetlands, and every year a million new houses and ten thousand new miles of asphalt encase more natural ground. This isn't happening only in the United States, of course. Since 1950, the paved surface area in the European Union has grown 78 percent while the population has grown 33 percent, and in the ever-expanding cities of the developing world, the trend is true as well. As one recent study began, "Paved surfaces are quite possibly the most ubiquitous structures created by humans."

As our cities and suburbs and small towns expand, a seemingly inexorable spread of pavement on which we walk, build, and live separates us from the natural ground itself, concealing from us our profound relationship not only to the source of our food, water, and energy, but to the many intangible ways the ground sustains our lives. Remarkably, many children now growing up in cities around the world rarely stand on unpaved ground, and fewer still ever stand on ground we might think of as wild. For anyone living in an urban area—already some 50 percent of the human population, and by 2050 almost 70 percent—it can sometimes feel as though the natural ground on which everything we know and love is built is itself disappearing or has already disappeared.

Of course, we need paved surfaces. The network of roads connecting our cities and countries allows dramatic and dynamic freedom of movement, and concrete literally supports our daily lives, from the foundations and floors of our houses and buildings to the bridges, walls, and infrastructures all around. In the end, the question we might ask is not whether we will have concrete and pavement, but how and where, and at what cost?

Soil scientists increasingly use the term "soil sealing" to describe what we are doing to our ground; not only are we paving more of it, but often we are paving our most fertile ground, eliminating its life, an act that is practically irreversible. Once we cover the ground with concrete or asphalt, they warn, we have lost it; practically speaking, that ground is dead. And this at a time when we have more and more mouths to feed and much less fertile ground than perhaps we believe.

I first understood this in the city of Parma, at the center of Italy's agricultural breadbasket, when an Italian soil scientist showed me an apple, compared it to the planet, then sliced the fruit in four. He discarded three of the four pieces as water (71 percent of Earth's surface), and kept the last quarter as representative of the continents. But this piece he cut in two, keeping only one, saying we discard the other because it is too rocky, or covered by ice or mountains. He then cut this remaining piece into four parts, and of these discarded three as areas too hot, too low in fertility, too salty, or too wet. "And so there remains this one small piece," he told me, holding it with his fingertips. He then peeled this piece, explained that the peel represented the thin layer of living soil over lifeless rock, and told me, "With this we have to sustain and feed the planet."

Ciro Gardi, the soil scientist with the apple, was just one of the people I visited while I researched this book. I spoke with many who know the ground well—geographers, ecologists, farmers, biologists, hunters, historians, and others—and asked them to look down and tell me what they saw. I wanted to learn the science of the ground and to find out how our changing relationship with the ground impacts our physical, psychological, and spiritual well-being.

I visited places where humans have a distinct relationship to the ground, moving from some of our oldest cities to some of our last wilderness. What's it like, I wondered, to climb down into a Roman cemetery in London, to share the sandhills in Nebraska with one hundred thousand cranes, to stand in solitude amid the parklike forest clearing in eastern Poland where nine hundred thousand were murdered, to lie back on the Alaskan tundra miles from the nearest road? The oldest traditions and the newest science both tell us the ground is alive in ways that we have never known or have long forgotten. I wanted to know how.

Through my travels I came to know the ground intimately, and to distinguish between different kinds of ground. These different grounds are reflected in the book's three movements: "Paved and Hallowed," "Farmed and Wild," and "Hell and Sacred." While these section titles reflect my journeys from one ground to another, I use the word "and" with purpose: the boundaries between these grounds often blurred.

In the first section, I start with some of our most paved places, the cities where increasing numbers of us live, and move outward into areas only recently paved or perhaps soon to be, ending on some of the most famous grounds of the American Civil War. This section has a deceptively simple ambition to help us become aware of the world beneath the pavement that is so often beneath our feet.

In the second, I focus first on soil, the life in the ground on which all other terrestrial life depends. Moving from grounds we farm to feed our bodies to wilder grounds that feed our spirit too, this section has as its goal an appreciation of the countless ways the ground sustains us.

In the third, I consider what might be hell on Earth, and then visit places that approach the sacred, reaching some of the most evocative and emotionally powerful grounds in the world. I wonder now about action: Knowing what we do about our countless connections, about destruction and care, what relationship with the ground will we choose?

I begin near the top of One World Trade Center. To the east across the water, a garden of towers spreads, marking Brooklyn and beyond. To the south, New York Harbor dominates. West across the Hudson, towers mark the Jersey City waterfront, then concrete to the horizon. It's not really true, as the advertising for the observatory claims with its trademarked slogan, that you can "See Forever." But I'm as interested in what can't be seen. Looking down at the countless buildings, the famous streets and bridges, only a few scattered patches of green or brown peek through, a few rows or groves of trees. The rooftops and avenues, the plazas and streets, all the surfaces are paved with asphalt and concrete.

Think of the sheer weight of all this civilization—buildings, pavement, people, cars. One recent estimate for the total weight of the buildings on Manhattan alone used the Empire State Building (350,000 tons) as a starting point and finished with a total figure of more than 118 million tons. The more than five hundred miles of streets and a similar number for sidewalks weighed in at more than four million tons. Add to that figure two million tons of vehicles (including 450 dump trucks at 35 tons each and more than six thousand 38-ton subway cars), pets, food, and people (which, added together, weigh about what the Empire State Building weighs), and the island supports more than 125 million tons of human stuff. The city, like cities everywhere, literally weighs on the ground, ground that is essentially entombed.

In the northern Minnesota lake country where my family has a cabin, it's tradition among the Ojibwe to begin any journey by praying in the four directions—a way of honoring and giving thanks. As I circle the 102nd floor, it feels like I am doing something similar. The view from here offers a chance for each of us to hold our breath "in the presence of this continent." The view from so far up reminds us that even when we can see "forever," there is more to see beyond and beneath, and more to know of a world that still calls for all our capacity to wonder.

PAVED AND HALLOWED

The ground we forget, and the grounds we remember.

Manhattan

ground[1] |
noun
1 the solid surface of the earth.
2 (grounds) the basis for belief, action, or argument.
— MERRIAM-WEBSTER'S COLLEGIATE DICTIONARY,
ELEVENTH EDITION

Next to One World Trade Center, as part of the National September 11 Memorial, are the footprints where the two buildings stood, square depressions, water falling down the four sides and flowing into smaller squares. At ground level, the chest-high walls are etched with the names of those who died. Maybe most affecting are the names of the firefighters, those who rushed to help others, their names grouped together as they were in life. Next to the building footprints, a plaza holds some four hundred swamp white oaks and several rectangles of turfgrass. Small signs warn visitors off the grass, and when I ask a guard why, she replies, "It's not a park; it's a place to pay respects." So I will follow a biblical lead. "Remove the sandals from your feet," Moses is told before approaching the burning bush, "for the place on which you are standing is holy ground." I take a seat away from the crowds, remove my shoes, and place my bare feet on the concrete. Sometimes in the northern forest, I'll walk the loamy ground under Norway pines, salmon-colored needles cushioning my feet. But in day-to-day city and suburban life?

The experience of feeling our feet on natural ground, common throughout most of human history, has become increasingly rare. And for anyone in a city, rightly so—the hard and often filthy artificial surfaces beneath our feet call for some protection. The skin on the soles of our feet is our body's thickest and some of its most sensitive, covering a high concentration of nerve endings. Our feet, in fact, are as sensitive as our hands—imagine wearing gloves as much as we have our feet in shoes. And the prints of our toes? As unique as our fingers—some militaries keep inked footprints as backup IDs—and the ridges and curves on the soles of our feet are as unique as those on the palms of our hands.

Around the world we have footprints preserved, in wet sidewalk concrete and clay Mother's Day gifts, but ancient footprints, too. Most recently a 1.5-million-year-old set was found in northern Kenya, likely made by a human ancestor called *Homo erectus.* So similar to the prints of a modern human are the fossilized find that the archeologist who discovered them said they look "like something that you yourself could have made 20 minutes earlier in some kind of wet sediment just next to the site." Wherever you are, a footprint offers an invitation to imagine those who walked here before you.

We are made for walking; our bodies evolved to perform this skill. And walking is maybe our most elemental experience of the earth, our most direct contact with the planet on which we rely. Walking (or bipedal locomotion, as it's known in some scientific literature) propels us at the perfect speed to soak up our surroundings, to talk with a friend, to contemplate the impending decision between a slice of pizza or lunch of some other kind. If travel by airplane erases the history, diversity, and beauty of the land we fly over, and driving reduces the world's sensory buffet to a tasteless, scentless, soundless blur, walking is where we have a chance to make it all up.

Maybe that's why so many of us claim that our best thinking comes while walking. Charles Darwin had a "thinking path" outside his home, a circuit on which he knocked a small stone from a pile each time around. Virginia Woolf used walking to get out of her house and venture into an exhilarating evening where women weren't often allowed, to "walk all over London; and see people and imagine their lives." Thoreau thought walking four hours a day to be living well, and wrote an essay in which he extolled the virtues of sauntering, to which the secret of success was "having no particular home, but [being] equally at home everywhere." William Wordsworth is said to have walked some 180,000 miles in his lifetime, which averages to more than six miles a day beginning at age five. There are countless famous examples of how walking stirs the soul and makes us who we are.

These days, though, we don't walk much. In fact, we in the United States walk less than the inhabitants of any other industrialized country. That my immediate examples above all hail from the nineteenth century makes sense, as even the "civilized" world was still made for walking then, not yet rearranged for cars, trucks, and buses. To mourn the loss of walking isn't merely to pine for the good old days, but to call attention to something we have always enjoyed. In New York, Paris, and other walking-friendly cities, we humans take advantage; we walk. I think we understand, as Rebecca Solnit explains in *Wanderlust*, that with the loss of walking we lose "an ancient and profound relationship with the body, world, and imagination."

We are learning more all the time about the true costs of this loss. As one *Slate* writer says, "The decline of walking has become a full-blown public health nightmare." One new study after another confirms that walking promotes health and well-being. This is true whether said studies claim that walking will bring "substantial reduction in the incidence of cardiovascular events," or that in terms of improving health, walking is "the equivalent of popping a series of magic pills." In addition to cutting our risk of becoming diabetic, suffering a stroke, or developing cancer, walking reduces hypertension, helps prevent depression, and strengthens the circulatory system. While Amish men take about eighteen thousand steps per day and Amish women fourteen thousand, the average American takes four thousand. Perhaps as a result, only 4 percent of Amish are obese, compared with 30 percent of the general population. Declares the nation's leading newspaper, "To Age Well, Walk."

At the southern end of Manhattan the streets still generally follow the more natural routes adopted by the Dutch from the Indians and developed in a way that reflected the island's hilly topography (the native Lenape name translates as "island of many hills") and swampy ground. Though it runs only as far north as Wall Street (once an actual wall built to keep the English or Indians out), the street map of the Dutch city (New Amsterdam until 1664) remains largely intact. Beyond that boundary lay wilderness. Here, across Houston, I find a very small park, just twenty-five by forty feet, that Time Landscape created in 1978 to preserve (or, actually, to re-create) the forest that "once blanketed Manhattan island." The surrounding area, says the sign, was "once a marshland dotted with sandy hills," the "trout-filled Minetta Brook" making it a popular site for fishing and hunting ducks. I stand outside the park's short fence on flat, smooth, hard concrete. On the other side is dark soil holding living things. Once there was a woodland here of witch hazel, black cherry, and red cedar trunks. Oak, sassafras, sweet-gum, and tulip trees, arrowwood and dogwood shrubs, bind-weed and catbrier vines, and violets. The small park invites city dwellers "including insects, birds, people, and other ani-mals" to experience "a bygone Manhattan."

Farther up the island, north on Fifth Avenue, the Empire State Building marks the horizon ahead, while behind me One World Trade Center marks the other. At Fourteenth Street I stop, for here begins a most remarkable story about the ground beneath this city.

In 1808, in an attempt to unravel the "evil of confused streets" and to facilitate the development of the island, city commissioners commissioned what would eventually become the street grid of horizontal and vertical lines that defines today's New York City. In doing so they turned to John Ran-del Jr., first for the grid's design — a survey that took two years to complete — and then for its execution. Turns out the survey was the easy part. To execute the plan, Randel had to mark the grid exactly (and this was a man of precision, in one case com-plaining that his measurements had set the grid off by 1.8 inches). For six years, the work took him through swamps and woods, farms and fields, over more than eleven thousand acres in total. He was pelted by landowners throwing artichokes and cabbages, chased off property by dogs, and arrested several times. Nonetheless, Randel and his employees placed more than a thousand bolts and monuments to mark future inter-sections. The development that would follow Randel's work took more than a hundred years to realize, but began almost immediately. Swamps were drained, woods were cleared, farmland and fields paved, the hilly island leveled—the bedrock just beneath the surface blown apart.

In fact, this drastic rearranging of surface life drew—even two hundred years ago—criticism for its brutish disregard of the island's natural topography. Walt Whitman and Frederick Law Olmsted criticized the plan, as did Alexis de Tocqueville, who claimed that it represented the imposition of "relentless monotony." Others criticized the chopping down of forests, burying of streams, and filling in of hollows to create a flat surface on which the commissioners could more easily lay out their grid. "The great principle which governs these plans," wrote one critic, "is to reduce the surface of the earth as nearly as possible to dead level." This is the reason Manhattan's streets are so flat and straight, its long rows of tall buildings creating canyons of stone that echo the sound of traffic, why twice a year the sun sets perfectly aligned along east-west streets, an event that's come to be called Manhattanhenge. Some ancient people's monuments to honor the sky and its gods? Or simply the unintentional result of what another contemporary critic called the "republican predilection for control and balance...(and) distrust of nature."

Though Randel and his men tromped through swamps, climbed hills, and crossed streams, almost none of those natural features figured in their final maps. In fact, even in their field notes the natural world was valuable only as a resource for human use. I think of historian Frederick W. Turner's description of European explorers discovering a New World filled with flowers and sea life and birds, a beauty worthy of inspiring song and poetry and praise—even "a new mythology"—and then his devastating line: "As it was, they took inventory."

Randel's maps take inventory as well, of ground that is portrayed as ready for the city's taking. In this way, Randel's hand-drawn and colored maps—which when connected to-gether to show the entire island stretch nine feet long—reveal the challenge posed by maps in general, that they are abstrac-tions of reality reflecting certain sets of values. It may be obvi-ous that no map can include every tree or bee or birdsong, but the implications may not be so. While on our maps we often see a park as a block of green, for example, or a lake as a blob of blue, more often a field of wildflowers or a stretch of woods is simply left blank. Even today our road maps—think Google—show only blocks of color dissected by streets. These are maps of abstraction, of time schedules telling us when we will arrive, not maps of living things. Whether we use them to understand the route through a city or to make decisions about what to "develop" and why, we would do well to pause. "Caution," a corner of each map might read: "No map can hold the true richness of the world."

Given that the Earth is spinning in space as it spins around the sun, and the sun is spinning in a galaxy that itself is hur-tling outward at a thousand kilometers per second, I some-times wonder how we are able to walk this ground at all. But as with so much else on this close-to-perfect world, it turns out that our planet spins at just the right speed to keep us happily walking crowded sidewalks, pizza in hand. The con-cept is this: as in a subway car or an airplane in flight, we seem set in place—we can walk around, stand on one leg, perform daring feats of balance—even though the train or plane is rushing us from one locale to another. It's only when that vehicle begins to slow or accelerate that we feel the pull—our body no longer quite sharing the vehicle's speed. All the spinning through space of celestial bodies doesn't affect us here on Earth because all these various bodies are moving at a constant rate.

There's also the fact of gravity. Newton gave us a basic understanding, which Einstein then complicated. Early in the twenty-first century, gravity is still acknowledged as "the most puzzling and least understood of the four fundamental forces of nature." Any simple definition of the word hides the fact that scientists still have loads of questions. "A force pulling together all matter," says one definition; another reveals that gravity not only dictates the movement of the planets in orbit around the sun and the movement of our oceans' tides, but also "the adherence of humans to Earth." And that's not all. Besides bringing such benefits as holding our atmosphere in place, gravity actually shapes our bodies. To put it another way, we are physically shaped — our bones and muscles molded — by our body's resistance to gravity's constant pull.

The mysteries of gravity relate to us in other ways too. After spending 340 days in space, American astronaut Scott Kelly admitted that the hardest challenge wasn't the effect of weightlessness on his body but "being isolated from people on the ground who are important to you." That pull toward our loved ones is a feeling we all can understand. And there's more: Kelly also felt a pull to the planet ("The *solid* earth!") on which we all rely. After witnessing great spreads of pollution, storms the size of which we have never seen, and the fragility of the atmosphere, he said, "I feel more like an environmentalist since I've been up here." Like the size of the universe or time before your time, gravity, a force we only begin to understand though it shapes Earth's life, seems a mystery too big to hold, and to which the only worthy response is astonishment and thanks.

*   *   *

Walking past Rockefeller Center I think of the annual Christmas tree in the square, sawed at its base and delivered by semi, set atop a sea of concrete. Everywhere in this city you will find that juxtaposition of the natural world with the human-made: the rooftop garden, the small pine on a patio thirty floors up, the overgrown rail line turned park. It's a juxtaposition dramatically displayed on the cover of Eric Sanderson's book *Mannahatta*—an aerial photograph of Manhattan taken from the southern tip, the right half of the island as it looks today, with its ground hidden by skyscrapers, and the left half made to look as it did in 1609, when Henry Hudson arrived, the ground hidden by trees, bursting green. It's this image that originally led me to contact Sanderson and ask whether we could meet, this side-by-side view of past and present. What I hadn't realized was that for Sanderson the image speaks of the future, of what could be—of what will have to be, if we are to thrive.

Wearing a white beard and a red stocking cap, Sanderson laughs as we meet, extending his hand. He'd brought along a copy of his book but left it on the subway. "So *Mannahatta* is riding around below us right now," he says. That which is below us has always fascinated him, including Umpire Rock, the outcropping on which we stand. "I feel like this rock is the symbolic heart of Manhattan," he says. It's remarkable rock, Manhattan schist, formed 450 million years ago, incredibly strong, gray, flecked with sparkling mica, its close-to-the-surface presence supporting the clusters of skyscrapers in Downtown and Midtown, and today the perfect hard sliding slope for wobbly little kids in neon down jackets.

"One of the things I like to do is to walk around in Manhattan," Sanderson says. "Where there are excavations for new buildings, I like to look in the hole, see what's down under the ground." It's ground, he explains, that is remarkably diverse, with some eighty-seven different kinds of soil within the city limits. "I see other people doing the same thing. I think there's huge fascination about what's underneath the surface, particularly in heavily developed places like Manhattan. I have a friend here in the Bronx who said when he was nine he was walking with his mom somewhere, and they saw a road crew working on the street. They'd dug down into the street, and he could see the concrete of the sidewalk, then the bricks underneath that had been there before, and then the soil. He said until that moment, he'd thought maybe it was just concrete to the center of the Earth."

Reminding us of the world beneath—and beyond—our pavement forms a critical part of Sanderson's work. Of his employer, the Wildlife Conservation Society, he says, "That's our goal, to connect the city and the wildlife." For a man whose work places him at the forefront of what can be a very discouraging fight to save wildlife and wild places, Sanderson is remarkably optimistic about being able to make that connection. "My feeling is that nobody really wants to destroy the environment. They just feel like they're forced into these decisions that they know are not so great."

Sanderson sees a desire for other options as reason for hope, reason for trying to figure out "how to talk to city people about wildlife and nature and wild places." People want to do the right thing, he believes, they just need to know they can make a difference. "There's that old quote about how you need to know something in order to appreciate it in order to love it and take care of it," he says. His wonderful book about the world under our feet was a major attempt to facilitate that process. "When I first moved to New York I was really struck how

New Yorkers love the great details of their block. *Mannahatta* is a way to tell them something brand-new about the place they lived that they never even guessed, and in huge detail."

And how to share that detail? In a word, maps.

Sanderson has been able to *show* New Yorkers—and all of us—something new about the place where so many of us live. By creating maps of the past and present, he offers us an understanding upon which to build our lives. "Can we travel back in time to see Mannahatta as it was?" he asks in his book's early pages. "And will that view change how we see our world today?"

Key to his mission are two complementary maps called *The Human Footprint* and *The Last of the Wild*, both of which visualize the human impact on the ground. Using data detailing population density, land use, roads, and artificial lights, Sanderson and his team of colleagues were able to show that at the end of the twentieth century, humans "directly influenced" 83 percent of Earth's land surface. The quotation marks are important, for if we factor in indirect influence like climate change and chemical pollution, that number reaches 100 percent. We have been—we are—nearly everywhere.

Critically, we take for ourselves huge portions of Earth's resources—a trend that threatens to increase as populations around the world move toward consuming at current developed-world levels. This spreading ourselves everywhere has meant devastation for our fellow creatures. In its most recent "Living Planet Report," the World Wildlife Fund declared that wildlife populations around the world have fallen by nearly 60 percent since 1970. This includes populations of mammals, birds, fish, amphibians, and reptiles, impacted by an ever-increasing human population that is clearing land to plant crops and expand cities. Making the situation even more dramatic is how rapidly these changes are happening. "Wildlife is disappearing within our lifetimes," said the group's director general, "at an unprecedented rate."

But this consumption of Earth appears to have serious implications for our own species as well. For example, Sanderson's team found that humans already influence fully 98 percent of the places where it's possible for us to grow rice, corn, or wheat—in other words, we already impact pretty much all the land on which we can grow food. This matters because by 2050 we will have three billion more people to feed. In fact, some estimates indicate that satisfying midcentury human demand will require the equivalent of four current planet Earths. Sanderson's maps are stark reminders that we have no new land to discover.

Especially for Americans, this can come as a shock. We have long based our culture on consuming what we know, then moving to new places. "I think that's an American conundrum," Sanderson says. "We haven't really spent a lot of time thinking about what we're giving to the future in terms of resources. We're very proud of this technology we've created and of the science that we've done. But we've been so wasteful of resources for so long."

At the same time, Sanderson's maps also show that the ground feels our influence on a gradient—in some places, such as cities, our footprint is heavy, while in other places it's relatively light. In fact, on 17 percent of Earth's land surface, that footprint is light enough that we might still call those places "wild."

Sanderson and I decide to walk to Times Square. Crossing over Fifty-Sixth Street on Sixth Avenue, he tells me that we have just walked up what was once a little hill. Looking back, I see the faint slope. But standing at this intersection once marked by Randel, that is the only sign of nature, this slight paved rise. "If we had the *Mannahatta* map we could tell we just walked across three ecosystems, three ecological communities," Sanderson tells me. We continue our walk down Sixth Avenue through the towering buildings. "It's not unlike the Grand Canyon," he says of the city. "It took a long time to build, you don't really know how it happened, it's huge!" Finally we reach Times Square. Walled in by buildings and bright flashing signs, we see no trees, no birds, no natural ground. It is, he says, "a temple of consumerism." And it's hard to imagine that nature was ever here.

Hard for me, maybe, but not for him. "There was a pond here, and streams that fed down what's now Sixth Avenue. People used to fish for brook trout here. It was famous for duck hunting in colonial times. This was a long way to come, three or four miles through wooded hills."

Once a wilderness north of town, this ground is now asphalt and concrete. Our footprint continues even under the surface, where below us run the arteries that keep this city pumping—power, cable, water, steam, and gas lines crowd the underground. And below those, tubes to carry the subway, and the sewage, and deep water lines, the deepest pipes reaching down eight hundred feet. No sign of soil, of green life emerging from dark underground. And no evidence that we are connected to that life. The illusion that we could exist without nature is complete.

Remarkably, the day I visit, one digital billboard advertises not a product but a project by the renowned architect Maya Lin, world famous for her design of the Vietnam Veterans Memorial. The ad celebrates Lin's "last memorial," called What Is Missing?, an interactive website featuring videos, audio, photos, and text that together present an elegy for lost and threatened species. As she says, "I am going to try to wake you up to things that are missing that you are not even aware are disappearing."

What is missing here in Times Square is almost everything. I remember when I first visited Manhattan some fifteen years ago I was shocked—coming from New Mexico—at the lack of natural life. I felt hemmed in and cut off; I wondered how people here survived. But since then, I've had more time here, enjoyed its parks, learned about the 250 species of birds that migrate through, and seen New Yorkers thrill to the wanderings of an adventurous coyote, to an eagle's unannounced stay, events that make front-page news. I have come to understand what Sanderson means when he talks about conservation—that it's not only about preserving those wild areas his maps identify, nature for nature's sake. "If conservation is also about conserving the human relationship to nature," he explains, "then this is the most important place."

The crucial question for conservation may be, Where to from here? It's a question that has special relevance for anyone living in a city, which means a large—and growing—percentage of humanity. Many of us will soon be living in megacities of ten million or more—already there are more than thirty, and by 2050 there will be some seventy-five—and many of these cities will be linked together into megaregions of continuous development. Sanderson is no Pollyanna, and he's realistic about what this means for the wild world. Human population growth, food production, land-use change, biodiversity loss—the impacts Sanderson included in the human footprint map—are going to grow more intense. Our human footprint is going to grow larger, and we are going to push the life of this planet harder than ever before.

Sanderson gazes at the throngs of shoppers and tourists, flashing lights and signs. "If we imagine nature after 2050—which isn't that far away—when seventy or eighty percent of the world's population lives in cities and the world population has leveled off at nine billion or ten billion people, what is that world going to look like for nature, and for people too?"

A fascinating question. And I know a city that might already offer an answer.

Mexico City

It's not what's on this island, but what's beneath it, that interests me.

— JULES VERNE, *JOURNEY TO THE CENTRE OF THE EARTH* (1864)

In what's known as the Valley of Mexico, concrete and asphalt flow from mountain slope to mountain slope, a sea made of pavement with countless buildings — houses, shacks, apartments, offices — riding the swells. Once, centuries ago, the valley was made of lakes, of fresh clear blues, including the most famous, Texcoco, where the Aztec empire was born. Those lake blues have vanished, their remains covered almost completely by a civilization made of asphalt and concrete. From an airplane window Mexico City is one vast spread of unending gray, accented with the ochre red of roofs and the dark pine of treetops, with only the rare patch of green ground to break the monotony.

Welcome to the quintessential megacity — defined as any urban area with ten million people or more. New York was the first, by 1950, and Tokyo followed soon after. Mexico City was the world's third, passing the mark in the mid-1970s. Over the past century the world's human population has grown exponentially (two billion in 1930, more than nine billion now), and the percentage of those people living in urban areas has grown along with it. As a result, the list of megacities had grown to twenty-eight by 2015; by 2030, Asia is projected to have at least thirty of its own.

To house, move, and support all these people will require a massive building boom, the likes of which the world has never seen. One gauge? Only twenty-five years ago, we were producing about three tons of concrete cement per year for every person on the planet; now that figure is closer to five. The explosive growth of Asian infrastructure leads the way—China alone makes and uses almost half the world's supply—but the use of concrete is growing everywhere. Because concrete cement makes up only about 20 percent of the total weight of concrete (important fact: though many of us use the words interchangeably, "cement" and "concrete" are not the same), the amount of concrete being laid every year is probably twenty-five tons for every person in the world. All this making of concrete has enormous ecological consequences, including producing 10 to 15 percent of our CO_2 emissions. It also says a lot about who we are as a species, and who we will be. "Concrete dominates the human environment," Robert Courland, author of *Concrete Planet*, told me. "Nothing is as representative of modern society as steel-reinforced concrete."

Mexico City is the perfect place to think about the consequences of all this paving and building. For one thing, under the combined weight of twenty-one million people and all their concrete, the city is rapidly sinking—over just the past half century an average total from locations around the city of more than thirty feet.

But the city is sinking not only because of what's on the surface, but also what's underneath. And what's underneath the pavement are the two stories that have brought me to the

Mexican capital. The first is a story of water, the second one of blood.

"Particularly in Mexico City we forget that we live within a lake," says Luis Zambrano, an ecologist at the Universidad Nacional Autónoma de México. "Everyone thinks that the lake doesn't exist anymore, and it exists." The lake Zambrano's talking about is Lake Texcoco, the site of the Aztecs' capital city of Tenochtitlán, the city the Spanish destroyed in 1519 before building their own city on top of it.

Founded by the Aztecs (or the México, as they are called here) early in the fourteenth century, Tenochtitlán began on a series of Lake Texcoco islands and expanded for nearly two hundred years as the Aztecs constructed artificial islands and canals for their buildings, plazas, and roads. When the Spanish conquistador Hernán Cortés and his soldiers conquered the city, they immediately set about destroying levees and dikes, draining the Aztecs' watery world—including Lake Texcoco—and building on the lake bed, sinking wells for drinking water into the aquifer below. That practice has continued in the centuries since, and as the Spanish city has grown in size, it has begun to sink into the old lakebed. On the city's Main Avenue, at the ornate Palacio de Bellas Artes, all white marble and pillars, visitors step down from street level before climbing the steps to the entrance, and doors that once were at street level are now several feet lower. The city's statue of El Ángel, the Angel of Independence, built on a hard sediment, has seemed to rise as the surrounding concrete around it has fallen—a century ago, nine steps led from street level to the statue; now there are twenty-three. In some areas of the sinking, city sewer pipes no longer flow outward, and sewage must be continually pumped to keep it from backing up. While the sinking can't be reversed, it could be slowed or even stopped, but to do so would take billions in investment and, perhaps even less likely, a systematic change in the way the city gets its water.

The problem is that modern Mexico City continues to take some 70 percent of its drinking water from the aquifers below its streets, and at an unsustainable rate. Twice as much water is being taken out as is going in, and scientists say the aquifer is now in danger of collapsing. Zambrano agrees, and worries that if the aquifer suddenly drops, the results could be catastrophic. "So for example," he tells me, "if in two years we have half the water or one-third the water we are consuming normally—and we are more than twenty million people— you can imagine that would create a huge problem."

I have come to the university's central campus, in the southern part of the city, to visit Zambrano at the Reserva Ecológica del Pedregal de San Ángel, the university's Biological Reserve, 237 hectares of native flora and fauna. A satellite view of the gray city shows two small patches of green at its southern edge. The first is the Biological Reserve. The second is Xochimilco (*so-chi-milco*), a UNESCO World Heritage Site, a wetland that is all that remains of the rivers and lakes that once served as the setting for Tenochtitlán. The aerial view tells a story of urban sprawl closing in on these last natural areas, the tide of pavement threatening to overwhelm the small remaining islands of what once was.

Standing in the way is Zambrano. Born and raised in Mexico City, Zambrano never imagined he would be where he is today. While around the world the other kids who would grow up to become ecologists played in the woods or fields or lakes near their homes, Zambrano grew up "surrounded by cars." He was originally drawn to study people rather than nature, and it wasn't until he started conducting research in Xochimilco that he began, he says, "to feel all the beauty; not understand it, feel it. To see a sunset, or in the early morning when you see all the birds, this is something spiritual."

As we walk the reserve, Zambrano tells me that even on the university campus, the seemingly inexorable spread of asphalt and concrete threatens this natural beauty. "We have to protect this all the time against everybody," he says. This despite the fact that already 98 percent of the city's natural ground is gone. "Now that I am in charge of the reserve, I tell everybody we should stop building. And they say no, it can't stop." He sees the city's failure to contain its growth as a failure to recognize its "carrying capacity," meaning the maximum population a given environment can sustain with food, water, habitat, and other resources.

"We are like an animal, an organism—and I am an ecologist, so I think all the time in these ways," Zambrano says with a smile. He explains that every organism, including a human being, has a carrying capacity. Humans are unique in that we can increase our carrying capacity with technology, but eventually, he says, one of two things will happen: "Either we will reach a carrying capacity that even with the most technology we cannot pass through and we will see that no, we can't grow anymore. That would be..." he pauses, "an optimistic outcome. The other outcome is that our carrying capacity has been expanded by technology a lot, and if the technology fails us, that carrying capacity will drop fast, and then there will be a lot of problems, a lot of deaths."

Despite what he admits is a sometimes fatalistic point of view, Zambrano is a warm host, full of enthusiasm. When he lectures to faculty and students, PowerPoint images of the city's growth over the decades draw gasps from his audience, while his regular jokes draw laughs. I see people nodding, listening, paying attention. He's appealing to common loves — slides of a baby fox, a hummingbird, bats. *"Muy bonito,"* he says of a particular flower, and everyone in the room agrees, it is very beautiful. With a photo of a resident frog he makes a *peep-peep-peep* sound. His dress shirt hanging untucked over his khakis, his hair somewhat mussed, his enthusiasm is infectious, as is his love for this small patch of native ground.

He asks rhetorically, Why protect this reserve? For *el paisaje*, the native landscape, he answers. For the native reptiles, animals, birds — mongoose, owl, salamander. The native plants — yuccas, agaves, and cacti, the complex native systems. *Dinámicas complejas.* For their sake, yes, and for how they affect Mexico's people too.

In the past, Zambrano believes, nature gave the various Mexican peoples their identities. Whether it was the Mayans (who thrived between AD 250 and 900), the Olmecs (who carved statues of giant heads from native stone), the Toltecs, or others, they knew who they were because of their relationship with nature. And their nature was specific; different areas had different ecosystems and species. Now, he says, pavement makes nature seem the same everywhere. Cut off from our heritage — the natural ground from which we came, a place our ancestors have known and loved and passed down to us as part of who we are — we lose a piece of ourselves.

In addition to his work here at the reserve, Zambrano's research focuses on the axolotl, a salamander clinging to existence in nearby Xochimilco. Zambrano tells me of the salamander's remarkable qualities of regeneration — its ability to regrow legs and spine and arms. But when I see them in his lab

I am stunned by their beauty. Each one is a small dark lion, its feathery gills encircling its head like a mane, and with a face that appears curved into a smile. Suddenly I see why the Aztecs saw mythic qualities in this salamander, recognizing in its regenerative ability a metaphor for the way the wetland ecosystem continually sustained their empire. For the Aztecs, axolotls would have been ubiquitous, and even as recently as 1998, the salamander seemed to be holding its own. Since that time, however, its population has crashed. For every sixty counted in 1998, Zambrano says, only one remains today. Their habitat has been drained and paved, their waters clogged with garbage, and an ill-conceived stocking of carp and tilapia has resulted in the destruction of axolotls' eggs. Today, perhaps only a hundred wild axolotls remain.

Zambrano sees the fate of the salamander and the fate of the Xochimilco ecosystem as inseparable. "It's like with a tree," he tells me. "If you cut down the tree, it won't be only the cutting of this tree but of the relationship between the tree and the nests of the birds, and the seeds, and the roots." And he sees the inability of his fellow citizens to restrain their city's constant growth — even at the cost of these last natural grounds, including species like this salamander that once symbolized their ancestors' civilization — as representative of an urban people cut off from the nature that sustains their life.

Intuitively, we might think, Yes, the environment matters, he explains, but the environment has become an abstraction; it's hard to connect with because we're not there anymore, and increasingly isolated from it. "The things that we should be concerned with are actually becoming more foreign and more unfamiliar to us," he says. This, he tells me, is the worst consequence of all the concrete.

I tell him of my first impression of Mexico City from the air, of how rare the green patches of nature seemed. Looking down from the plane I wondered how far children—or anyone—must have to travel to reach this green. Here in this paved valley, do growing minds imagine this is all there is: a world made by humans for humans as far as the eye can see? Do weeks or months or even whole childhoods pass without hands in the dirt, feet on the grass, ears filled with birdsong or cricket saw or moonlit quiet?

Zambrano nods, and then he tells me about taking his daughter up onto the roof of their house. "In Mexico, things are a little bit dangerous," he says, between the narco drug-trafficking violence and the impunity with which criminals operate. "At her age, I used to leave the house without any problem and go to the store or even take the bus. I mean, my mother knew I was safe. There were risks, but normal risks. Now, my daughter—she is twelve—wants a lot to go to the park that is in front of our house, and we don't let her do that because we are really concerned about security."

We are walking through the Biological Reserve, sur-rounded by the remnants of the region's native flora: prickly pear, cholla, yucca—some of Mexico's nearly one thousand species of cactus, the ground in places black, lava from ancient volcano eruptions.

"One day, last week," he continues, "it was a very nice evening, at sunset, and I told her, let's go to the roof because maybe we can see the volcano, because sometimes from the roof you can." Zambrano is speaking of Popocatépetl, which in the Aztec language means "smoking mountain." The beautiful snowcapped volcano, forty-some miles south, tow-ers over the city and—if the air pollution isn't bad or the clouds don't block it—serves as its backdrop. Zambrano tells me the climb to the roof isn't easy, that it's a bit dangerous, and that his daughter had never been there before. When they reached the roof, she smiled.

"She told me, this is nice, and she asked, 'Could I come to the roof sometimes? Because this will be my way of freeing myself. If I can't go to the park,' she said, 'then let me stay here.'"

Zambrano stops and looks into the distance. "And when she said that, it hit me," he says, "that as a human she needs to see scenery. It's not enough to be in the house. She is cut off from nature."

A few days later, in the Zócalo, the old city's center square, I myself am feeling cut off from nature. Surely one of the world's great demonstrations of paving-gone-crazy, the large square has zero green—no trees, no plants, no bushes—just a tall flagpole flying an enormous Mexican flag. This area has been a plaza for centuries, and during Aztec times it was the central market. Bernal Díaz del Castillo, a foot soldier in Cortés's invading army, wrote, "Among us there were soldiers who had been in many parts of the world, in Constantinople and all of Italy and Rome. Never had they seen a square that compared so well, so orderly and wide, and so full of people as that one." But around the square the Spanish soldiers also found evidence of human sacrifice, including a stench of burning human flesh so thick, wrote Díaz, the scene "brought to mind a Castilian slaughterhouse."

These days the square lies framed by stone government buildings, tourist hotels, and the enormous cathedral built by the Spanish from 1573 to 1813. It's the cathedral I have come to see—or rather, the ground beneath it, where in 1978 an electrical company crew accidentally discovered the center of the Aztec universe.

The story of the Aztec civilization has long captivated us. Their main city centered around this great square with two pyramids at one end, one of which was the main temple, the Templo Mayor. From 1325 to 1521, the Aztecs built an empire through military conquest and ever-expanding trade. But the empire grew increasingly top-heavy, expanding beyond its means, and by the time Cortés arrived, in 1518, the Aztecs were ripe for a fall. Over the next four hundred years, as the Spanish—and then Mexican—city grew, the old Aztec city lay underneath. Well into the twentieth century, archeologists didn't know the exact location of the great temple. But since its accidental discovery, more than fifty thousand Aztec objects have been unearthed, many of them exquisite statues and carvings. And the work continues; to visit the Templo Mayor today is to visit an active excavation site.

I visit to meet with Eduardo Matos Moctezuma, who since the 1978 discovery of the Templo has devoted his life to its cultivation. Friendly but reserved, Matos has an eminently authoritative air. In the on-site Museo del Templo Mayor, the wealth of artifacts around us is overwhelming—here a canine skeleton, there a box with thousands of seashells, around the corner an incredible ceramic figure of an eagle, another with the terra cotta shape of a bat (the "original Batman," he says, smiling). What's amazing is the way this prehistoric world has emerged from beneath such a solid layer of concrete. It makes you realize what you're walking over—the cultural history—as you walk Mexico City's streets. "To break this concrete wall," he says of the capital's paved surface, "you're going down five centuries." Once, he tells me, a reporter asked him how many archeological sites there are in Mexico, thinking there could be more than a hundred thousand. And Matos told him, "Don't worry, there's only one, and it's called Mexico."

The Templo Mayor was the Aztecs' most sacred ground, the place on Earth where the nine levels of the underworld and the nine levels of heaven met. (Anyone familiar with Dante's fourteenth-century *Divine Comedy* will know that in his *Purgatorio* he envisioned nine levels of hell, and in his *Paradiso* nine levels of heaven. Incredibly, across the world at the same time, the Aztec worldview was remarkably similar.) The Templo Mayor was, Matos explains, "the fundamental center of the Aztec universe, or their cosmovision. It was also the site where the vertical and the horizontal planes met, that is, the passage to the upper or celestial levels and to the underworld, and also the place from which the four directions of the universe departed." In short, "There is no doubt that the Mexica thus claimed to be the center of the universe, the chosen people."

That said, the Aztec practice that probably fascinates us most is the custom of human sacrifice. And how can you not be fascinated by accounts like this: "The most significant act in the feast was the sacrifice of a woman who was carried to the top of the temple back to back with a man and beheaded there. Afterward the man was dressed in the skin of the sacrificed woman and in the weaving she had done before her death." Why was human sacrifice so important for the Aztecs? We might be tempted to see it only as a barbaric act, a mindless waste. In fact, says Matos, "the main goal was to preserve life. You had to give life to preserve life. Most of the human sacrifices were to maintain the movement of the sun, on which all life depended. You had to give blood to the sun."

In the ritual of human sacrifice, captured soldiers, slaves, or others—often dressed to represent Aztec gods—were brought to the top of the Templo. Their chests were opened and their hearts torn out. (How does that even happen? An artist's rendering in *National Geographic* depicts the victim on his back on an altar, soldiers pulling tight on each arm and leg, bloody knives, and a priest holding the dripping heart high over the blood-soaked chest.) Their bodies were then flung down the Templo steps to the plaza below to be dismembered. Matos explains that the Aztecs made human sacrifices to the gods "so that the sun would not stop its course. The gods had to continue to die in order for the universe to continue. A man, converted into a god at the moment of sacrifice, performed what the god did in *illo tempore*."

I have to ponder this, imagining those who were to be sacrificed climbing the Templo stairs. Did they understand the role they played? "Absolutely," he says.

We in the modern West have a hard time grasping the idea of human sacrifice, wanting to consign it to legend, thinking, How could this kind of ritual death happen on such a big scale? But there's no doubt that these rituals took place. Recent tests of porous surfaces in Mexico City have revealed "blood traces everywhere," says one archeologist. "You have the sacrificial stones, the sacrificial knives, the bodies of 127 victims—you can't deny the human sacrifice."

And "it should be pointed out," Matos writes in his *Life and Death in the Templo Mayor*, "that this kind of thinking derived from humans' observation of everything around them: people, plants, animals, the day. Everything is born and dies, a cycle that is constantly repeated. It is the only way the universe and its renewal can be conceptualized. Everything has to go through the cycle, so there is no belief that people or the sun to which they belong will endure. That would break the pattern of everything created. What has to be maintained is the constant cycle."

Perhaps sensing my difficulty grappling with this practice, Matos says, "I will say something now that perhaps you do not want to hear. What about your wars in this century, the Germans killing six million Jews, the Americans killing one hundred thousand Japanese with one bomb?" He allows the question to sink in, then says, "And that's not for life to go on, but for death."

Paved over often for centuries, this too is something we find in the ground.

London

*It is wrong to think that bodily health is compatible with
spiritual confusion or cultural disorder, or with polluted air
and water or impoverished soil.*

— Wendell Berry, "The Body and the Earth" (1977)

W hat you're looking at there," says Nick Eldsen, "is
people standing waist-deep in the medieval period."
I am with Eldsen, an archeologist with the Museum of London, in an office building that until recently looked down on
one of the busiest streets in London, a block from Liverpool
Street Station. But now this building looks down on an enormous pit—what will become an underground train station,
and what was once a medieval burial ground. In other words,
we are looking down into the past, and in a few minutes we
will be there ourselves.

The site is a future station for Crossrail, an incredibly
ambitious project to run a train east to west under the city.
The project involves ten thousand workers, forty construction sites, and twenty-six miles of tunnels dug deep below
London's streets—and between, over, and around the network of pipes, tubes, and various utility lines already creating a complex maze beneath the city's surface. This particular
tunnel will run within two feet of the Underground's Ham-
mersmith and City Line, one foot under the Post Office

Railway tunnel, and close under the Goswell Sewer, the second-largest sewer in the city. In the room behind us and on several floors above and below, hundreds of engineers and project managers stare into computer screens, huddle in conferences, and explain into phones, each with a bright orange construction site jacket hanging from their chair, back, or coatrack. I have donned the same orange coat plus orange overalls, rubber boots, an orange helmet, plastic glasses, and gloves. Wherever we go, there's no doubt we will be seen.

A longtime archeologist for the museum, Eldsen has worked on dig sites all over the city. But the chance to peer into the ground under one of London's busiest streets is a rare opportunity indeed, he says. The Crossrail project has created the "largest archeological dig in London for decades," one that has yielded bones from reindeer, bison, and mammoths that date back nearly seventy thousand years, an eight-hundred-year-old section of ship, and the remains of Tudor houses. Among the spectacular finds on this particular site near Liverpool station are a graveyard populated by former residents of the Bedlam Hospital, with skeletons sometimes packed eight bodies per cubic meter, coffins laid "head to tail like sardines in a can," Eldsen tells me, "and half a cat."

"Half a cat?"

"Yeah. Also, there was a little girl that had been buried with beads. And a woman with a plate, a seventeenth-century Delft blue-and-white, buried upside down on her chest, probably her favorite." As I am imagining this woman with her plate lying buried for nearly four hundred years just a few feet below the surface, Eldsen says, "Millions of people must have gone walking through here on Liverpool Street, a busy road leading up to a mainline station, and not had any inkling what was underneath."

It's these skeletons that have "a bit of personality" to them that make an archeologist's curiosity come alive, Eldsen explains. "I spent years on a medieval site by the Tower of London where we found Roman burials with the heads cut off, and we don't know why." In another dig, archeologists discovered a medieval school where under the hallway floor they found a wooden shoebox, and in it a (whole) cat, wrapped in a shroud, perhaps by Tudor law students sometime between 1400 and 1600. Eldsen admits that such finds are rare, but "every now and then you do get something that allows you to get a view of a little event going on."

That is certainly true at the site we visit now. Having come down into the street to stride past busy Londoners in our swishy orange pants and awkwardly fitting hard hats, we descend by aluminum ladder into the actual pit. The first five meters hold most of the archeological finds, Eldsen says. Once you get past about ten meters down, there is no more evidence of the human. (And think of that! All this time on Earth, even here in one of the most intensely lived-in locations on the planet, and our history is only thirty feet deep. "You go down deep enough and the human story ends?" I ask another archeologist. "The party's over," he says with a nod. "We call it subsoil.") Around me an archeology team is digging carefully and brushing dirt from ruins. A defunct nineteenth-century brick sewer pipe runs nearby, filled with concrete so the surrounding work won't collapse it. A few feet away sit several large clear plastic bags full of bones. Eldsen reaches into the closest and pulls out an enormous piece, which he nonchalantly identifies as a cow's jawbone. "It's all really just cooking waste," he says.

Before the archeologists can dig back into the past, says Eldsen, the construction team must cut through the modern street surface, a process that sometimes can be quite loud. "I swear that I can hear the opening rhythm of Jimi Hendrix's 'All Along the Watchtower' from these things when they're going full tilt," Eldsen says of the street-cutting machines. Sometimes the process will uncover the stone "setts," or cobblestones, of a Victorian road surface, but most often they find the first layers of rubble and dirt that were laid down between phases of building from the Roman period to the present day. The archeologists rely on careful machine sifting, a process that stops whenever some more significant features start to turn up, such as the walls of a seventeenth- or eighteenth-century building poking up through the Victorian layers sealing them. At that point the full archeological team jumps into action, cleaning up by hand with mattock, shovel, and towel, before they start the systematic process of recording and removing each relic in turn.

Here on Liverpool Street what has especially inspired Eldsen and his colleagues is the discovery of sixteenth-century burial grounds. Just a "football-pitch-sized piece of ground" holds more than two hundred years of burials and more than twenty thousand skeletons packed tightly together. Eldsen and his colleagues are eager to find out what killed them. "If this isn't plague, it's smallpox or another epidemic. Most people died fairly quickly, and so that doesn't impact the bones. What you get is what they lived with—signs of rickets, bowed legs. There was one poor lady who had syphilis that had eaten into her bones. Lots of bones that had broken and not healed straight. So there's an amazing amount of information in here."

An amazing amount of information, agreed. And maybe it's knowing that so many have lived, died, and been buried near this spot. Or maybe it's the image of the skeleton holding her plate just a few feet below the sidewalk. But for me, peering into the past with Eldsen gets me thinking about the relationship between the ground and our health. It's a relationship, it turns out, as intimate as they come.

My first stop after meeting Eldsen is to lunch with Dr. Graham Rook. I have wanted to speak with Rook ever since discovering his "Old Friends" hypothesis—that our living in urban areas, separated from the microbiota in the natural ground, is having serious consequences for our health, and especially for our children's health. While an ever-increasing number of studies have built the case that living closer to "green space" is beneficial, Rook's work identified one of the direct reasons why. I find him jovial and friendly, with an amused take on the world.

"We now know that our immune systems, when we're born, are like a computer with a hard disk but no data, and without data it doesn't function," he tells me in a cafeteria near his office. The most important thing is that the data teaches it to regulate itself, he explains, because the immune system is extremely dangerous. If it turns on you, it will kill you. And, in fact, in modern cities we are developing more of the disorders in which the immune system is not correctly regulated. For example, we have more diseases like multiple sclerosis, in which the immune system is attacking our own tissues. And we have more people with inflammatory bowel diseases, where the immune system is attacking the vast numbers of organisms in our gut. Counterintuitively, these attacks don't occur in developing countries, or they are very rare. So, Rook argues, we need to have the input in the immune system in order for it to not attack the things it shouldn't be attacking. "What we're getting in the rich countries, especially in urban populations, is increasing incidences of all these diseases. It's not the only reason why we get these conditions, but it is the major contributing factor."

By way of example, Rook tells me about a pediatric epidemiologist named Erika von Mutius who investigated asthma in East Germany after the Berlin Wall came down in 1989. Assuming that there would be massive levels of asthma because of all the industrial pollution, she was surprised to find very low levels. "We now know that pollution and tobacco smoke don't cause the asthma, they trigger the symptoms of people who have it. Who are the people who'll have it? Well, they are people who don't have their immune system correctly regulated."

Another example? "Hay fever started to be noticed in the early nineteenth century. A chap called John Bostock, an Englishman, in 1914 he got very excited about it. He traveled all around England looking for cases. He found about twenty cases, having, he said, scoured the whole of England. I bet you I could find forty cases in here today," Rook says, scanning the crowded room. "Something has changed."

The bottom line for Rook is that "if you meet the microorganisms that are present in a cowshed in the first two years of life, or if your mother meets them while she is carrying you as a baby, then you are less likely to have allergic disorders, you are less likely to have inflammatory bowel diseases in childhood. And you're almost certainly less likely to develop psychiatric disorders as an adult as well." Researchers in Finland, he says, have found similar results.

"The only good thing Stalin did—from an epidemiological point of view—was when he invaded Karelia," Rook explains. Ever since, Karelia has been divided between Finland and Russia. While the Finns live in well-insulated houses and have a high incidence of autoimmune diseases, such as Type 1 diabetes and childhood allergies, the Russian Karelians are still practically living in mud huts and have very few of these disorders. Finnish researchers have compared the microbiota in the house dust in the Finnish houses and the Russian houses and found them to be completely different. In the Russian houses there are lots of animal-derived strains, lots of environmental strains, "lots of things wafting in from the soil and the environment," Rook adds. The Finns called the connection between microbiota and human health the "biodiversity hypothesis."

Whether we call it the Old Friends hypothesis or the biodiversity hypothesis, mounting evidence suggests that our obsession with hygiene and "cleanliness" in the West, and our associated aversion to dirt, is having unhealthy results. Rook says this aversion to dirt is a misunderstanding of the human place within nature. "People tend to assume that humans are some kind of plastic spaceman that arrived here and has nothing to do with this biosphere," he says. "Of course that is complete nonsense. We are a part of the biosphere, and we are ourselves an organism that overlaps with and interacts with the ecosystem."

The notion that the human body is separate from the natural world is an old one, and it has warped our thinking throughout the ages. For example, says Rook, "People had always assumed that your lungs are sterile, which, when you think about it, only the medical profession would be quite so stupid. They are our core connection to the outside world. We're a bloody great tube," he says, laughing. Rook explains that "there's all sorts of stuff in the airway, and there needs to be." In fact, the number of organisms in the air is stunning, he says, and the complexity of the microbiota of the air is almost as complex as that of the soil. Most of it is in the form of particles, which will then be picked up by the cilia ("the little hairy things that wave inside the airways," he explains), balled up, and then swallowed. Indeed, most of what we breathe in will ultimately end up in our gut.

Is Rook saying we have lost contact with the soil and that is having these effects? "I'm saying we've lost contact with the green environment. I regard soil as very much a part of that. Because most of the organisms that are wafting around in the breeze—I love that word, 'waft'—are in fact organisms from the soil."

Have we really lost contact with the green environment? It may not seem so. After all, many of us can see trees and grass out the window, perhaps a squirrel or some kind of bird. But when I think about the last time my bare hands actually grasped fistfuls of soil, or the last time my bare feet felt the natural ground—a sandy beach, pine needles in the forest, a front yard's fresh-cut grass—I think, *Yes indeed, these experiences are becoming more rare.* That word "contact" is key. The "state or condition of physical touching," says Merriam-Webster, and I think again of Henry David Thoreau on Maine's Mount Katahdin, climbing the mountainside alone in mist and fog, feeling like the first person on this part of Earth—and then feeling his connection to physical ground.

It's easy to imagine Thoreau grasping at his chest, his arms, his face—feeling a human body that is like the body of the solid earth, in fact a part of it, not separate from it. In *Walden*, he asks, "Am I not part vegetable mold myself?" That we are part of the natural world, part "vegetable mold" ourselves, is among Thoreau's many radical insights, and perhaps the most challenging to our modern senses.

*   *   *

A rapidly increasing number of studies explore the implications of this connection for our physical, mental, and even spiritual health. While direct causation between exposure to nature and improved human health is sometimes difficult to prove, associations between the two are not. In fact, again and again, reports indicate that "the balance of evidence indicates conclusively that knowing and experiencing nature makes us generally happier, healthier people."

Some may say, Well, of course. Or, Why would we need scientific studies to verify what everyone knows is true? Unfortunately, in cities around the world, more and more of us are living without the opportunity to experience nature. And, remarkably, even in Western industrialized nations like ours, it is especially true for children.

Probably no one in recent memory has more successfully raised awareness about this problem than Richard Louv, who called attention to what he called "nature-deficit disorder." He wrote,

> Some startling facts: By the 1990s the radius around the home where children were allowed to roam on their own had shrunk to a ninth of what it had been in 1970. Today, average eight-year-olds are better able to identify cartoon characters than native species, such as beetles and oak trees, in their own community. The rate at which doctors prescribe antidepressants to children has doubled in the last five years, and recent studies show that too much computer use spells trouble for the developing mind. Nature-deficit disorder is not a medical condition; it is a description of the human costs of alienation from nature.

One striking manifestation of this change was the fact that in 2008 the publishers of the *Oxford Junior Dictionary* replaced dozens of nature-related words with more tech-oriented words. This action struck so many people as too impossible to be true that the fact-checking website Snopes .com investigated. Their verdict? It was true: the dictionary had been "systematically...stripped (of) many words associated with nature and the countryside." The publisher defended itself by arguing that the omitted words no longer hold relevance for modern-day childhood, and that they had only so much space ("little hands must be able to handle it"). In 2015, a group of writers led by Margaret Atwood asked for reinstatement of the words, citing the unraveling of the bond between nature and culture "to the detriment of society, culture, and the natural environment," but to no avail. Among the words removed: acorn, dandelion, fern, heron, lark, otter, pasture, and willow.

The word-stripping may seem less startling when we consider the actual amount of time children spend in nature these days. Recent studies report that American children have "on average only four to seven minutes of unstructured outdoor play per day." And, American children spend "more than 30 hours per week connected to electronic devices, but less than an hour a month in nature." Said one study, "Children are in some ways on house arrest." Another described today's kids as "the backseat generation" whose "experience of nature most often occurs from the inside of an automobile." Yet another report reminds me of Luis Zambrano's daughter in Mexico City: "82 percent of mothers with children between the ages of 3 and 12 cited crime and safety concerns as one of the primary reasons they do not allow their children to play outdoors." The last phrase of this sentence might sound like a typo: "do not allow their children to play outdoors." But in fact, today's children spend less time playing outside than any generation before them.

We might be most troubled by the lack of contact with nature for our children, but anyone living in an urban area does so with less natural diversity than did preceding generations. "It is likely," one study reports, "that most of Earth's urban human population lives in biological poverty. In cities all over the world, there are fewer kinds of birds, flowers, insects, and other wildlife, as well as simply less of it overall." In fact, the authors argue, this holds true for "cities diverse in age, size, location, and surrounding habitats." And not surprisingly, the most severely affected are poorer and disadvantaged communities. Two examples: In 2013, the *Washington Post* found that "wealthier areas of the District of Colombia had an 81% average tree-cover rating, while lower-income areas averaged only 48% coverage." Another study found that "land cover was associated with segregation by racial and ethnic minorities. A problem that...raises serious issues of environmental justice."

In 1984, Harvard biologist E. O. Wilson coined a word to describe what he saw as an instinctual human need for contact with nature: "biophilia," meaning "love of life and living systems," or, as Wilson put it, "the connections that humans subconsciously seek with the rest of life." Wilson argued that biophilia is a biological need essential to optimal health, and his writing ("the brain of the fly...resembles a grain of sugar"; "every species is a magic well") is brimming with it. "To an extent still undervalued in philosophy and religion," he wrote, "our existence depends on this propensity, our spirit is woven from it, hope rises in its currents." Thirty years later the scientist would write of how our destruction of the world's biodiversity was leading us into the "Eremocene," the Age of Loneliness.

Study after study has supported this hypothesis, linking contact with nature to long- and short-term mental and physical health. A recent Danish review of academic studies found direct health benefits that "included psychological wellbeing, reduced obesity, reduced stress, self-perceived health, reduced headache, better mental health...reduced cardiovascular symptoms and reduced mortality from respiratory disorders."

Those of us living in urban areas have a much higher risk for anxiety, depression, and other mental illnesses than those who don't. At the same time, more of us than ever before in history spend our lives indoors and on concrete, literally separated from the natural ground, and spend less time than ever before in contact with natural green spaces.

If we suffer as we remove ourselves from nature and we prosper as we are reconnected to nature, it makes sense to ask why we continue to live as we do. Why we continue to pave our cities so completely. Why we undervalue open space. Why we allow the type of development that reduces our contact with nature. Especially when, as one researcher argues, "If we had a medication that did this—a medication that prolonged life, that addressed very different unconnected causes of disease, that did it at no cost and with no side effects—that would be the best medication of the decade. But we don't have a medication like that except for this 'vitamin N'—nature."

Unfortunately, says Canadian neuroscientist Colin Ellard, "Historically, the attitude toward the importance of green space has been basically to consider the presence of greenery as an esthetic nicety, rather than as something of fundamental importance to people's psychological state." Even now, the ever-increasing scientific evidence about how nature affects our health is only sporadically finding its way into our policy decisions, especially in terms of land use. As one recent report found, "Probably the one area where rapid progress could be made is improving communication and collaboration between land-use and city planners, (and) people involved in public health."

There's something else going on here too. A major hindrance to our stemming the tide of pavement rising across our lands is that it's often easier to see the immediate monetary benefits of such development than it is the longer-term negatives. My host at a recent astronomy talk complained of a new development that would "almost completely surround (his) observatory with light," but in the same breath acknowledged, "I have to admit that I'm for it, a position that may reflect my age—twenty-nine—and a desire for a robust economy during the majority of my career." It's probably far harder to see the costs of a "robust" economy—and by this, we most often mean an economy of endless growth—because they don't immediately impact us. If the loss of natural spaces—and our resulting loss of contact with them—often doesn't seem to have an immediate effect on us, how can we begin to truly recognize or quantify that loss?

That's a question Graeme Willis and Emma Marrington at the Campaign to Protect Rural England try to answer every day. On a foggy London morning, I visit the CPRE offices on a winding backstreet in South London. Graeme and

Emma welcome me warmly and share the ways CPRE is trying to protect what they see as a resource as vital to the country as any other, the actual countryside.

Created in 1926, CPRE has long had as its mission the
goal of shaping development in England so that the national
heritage of "a beautiful and thriving countryside that's val-
ued and enjoyed by everyone" can be preserved. The group
has had considerable success, helping to create several national
parks and influencing planning laws on local, regional, and
national scales. Recent campaigns include those designed to
protect England's historic hedgerows, slow the expansion of
roads and housing developments, and limit the damage created by new fracking operations. But the heart of their work
remains raising awareness that the English countryside is
under constant threat of reckless development.

For Graeme and Emma, the idea of our separation from
natural ground hits close to home. During the 1990s, they tell
me, CPRE often used the phrase "concreting over the countryside" as a way to capture attention in the popular press.
Even now, Emma says, "our current president, Sir Andrew
Motion, who's the former poet laureate here in England, talks
a lot about 'drooling concrete' over the countryside." Already
the most densely populated country in Europe, England continues to see a steady spread of pavement. To help the public
understand what's being lost, CPRE found a word that cap-
tures a quality they want to protect, and then they made a map
of that quality. A map of tranquility.

Tranquility, according to CPRE, "reflects the degree to
which human beings experience the environment unhin-
dered by disruptive noise, movement, and artificial lighting
and structures." The quality of tranquility is, they say, "one
of the countryside's greatest gifts to us all." The map, which

colors the country from deep green ("most tranquil") to deep red ("least tranquil"), with lighter green, yellow, and lighter red between, makes its point immediately—only a few areas of deepest green remain, mostly in the north, while the deepest red of urban areas spreads along roads and highways like a spider's web to nearly every area of the coast.

CPRE created the map by using information about the presence or absence of wildlife, open spaces, and natural landscapes, or roads, buildings, and lights—the paved world we know well. In fact, the areas that are the deepest red are the places where most people live. The map, Emma explains, is "a way of identifying, measuring, and mapping tranquility so that it can be integrated fully into public policy decisions." It's something the government has been slow to do. As official CPRE policy explains, "The overwhelming tendency over the last 30 years has been to fragment and obliterate tranquil places and reduce this quality where it is still present."

A big part of the problem is that, faced with new roads, buildings, and lights, qualities that contribute to tranquility such as quiet, darkness, and calm "are easily overwhelmed by the scale and power of such intrusions." These intrusions can start small and spread slowly, and by the time we realize what's been lost, it's too late. The result—many now argue—is that the developed world is home to more and more people who are sick in spirit, more and more people suffering from stress and mental illness.

But the map isn't simply about pointing out the reddest areas. This map, like Eric Sanderson's *The Last of the Wild*, is also about helping people see the tranquil areas that are left and trying to, as Emma says, "reconnect people with their local countryside." It's a countryside with which increasing numbers of Britons have lost touch—a fact of modern life that is different from just a generation or two ago. "Growing up, I was always out, getting muddy, rambling, so to me that was normal," Emma says. "Whereas a lot of kids nowadays don't have that sort of experience."

Says Graeme, "Something we've got to get to grips with is the notion that our best interests are served maintaining that severance, whereas actually they're best served by realizing we are in nature, we're of nature, it's in us." It can be hard to realize this in a world in which our needs are constantly being anticipated and catered to with new products. As Graeme puts it, "Somebody could say, 'Hey, we can make a lot of money producing a virtual-reality headset that takes people into a tranquil environment.' And to be fair, if you need to get away from it all, putting a headset on for an hour might actually do the job in some way. But the assumption would be that that's doing the whole job.... We might know that seeing natural environments and seeing landscape and hearing birds is what does it for us, but it might also be all these things combined."

I am struck by the way the CPRE's map of tranquility is in some respects a map of roads. In fact, especially when talking of paved roads, it's hard to overstate their effect on the planet. The ease of transport roads provide is a central pillar of our society. But when we follow our roads, we bring along that society with us, and oftentimes the not-so-tranquil aspects of ourselves.

Near the end of my visit to the Crossrail dig with Nick Eldsen, we got down to where, as he put it, "we get the Roman stuff going on." He was speaking of Londinium, the Roman city that existed here between about AD 43 and 400. Together we looked at what he called "a nice Roman road coming through," something he and his colleagues had figured out after finding quite a lot of Roman horseshoes. "They didn't work very well," he explained, "so there's a lot of them from this area, and we've actually got them in the wheel ruts in the road. And if you can imagine—we're certain there must have been a little bridge here—some carter struggling with his horse to get up the ramp to the bridge. And the horse loses his shoe in the mud, and it provides a nice little snapshot of Roman everyday working life."

In Britain, the Romans built one of their most extensive road networks, with routes that major highways follow today. The United Kingdom is also where paving technology developed fourteen centuries later, in the 1800s, technology that soon sprawled out around the world, paving the wild straight to our very front doors.

Northern Virginia

The costs of all this driving…are beyond calculation. The least understood cost—although probably the most keenly felt—has been the sacrifice of a sense of place: the idea that people and things exist in some sort of continuity, that we belong to the world physically and chronologically, and that we know where we are.

— James Howard Kunstler, *The Geography of Nowhere* (1993)

Just before Interstate 81 north out of Virginia's Shenandoah Valley crosses I-66 west from Washington, DC, I exit for the Cedar Creek Battlefield to meet with Mike Kehoe, "digger." Kehoe is other things, of course, including the retired town manager of nearby Stephens City and an incredibly nice guy with a thick northwest Virginia accent. But, as he tells me, "I've always had my nose in the ground ever since I was five or six years old." For as long as he can remember he has been digging artifacts from the local dirt, a popular hobby in a part of the country where the American Civil War raged for four long years. Bullets, belt buckles, bayonets—even now, more than 150 years after the war's end, the ground here holds undiscovered treasure.

At the same time, this area has in recent decades seen huge increases in its human population, and much of its former rural character has been paved away. For many people here, especially longtimers like Mike Kehoe, it's not so much the development in general that makes them uneasy, but a style of development "marked by automobile-dependent, spread-out suburbs, where the activities of daily life—home, school, shopping, and work—are separated by long distances linked only by pavement." In other words, a style made possible by paved roads known as "sprawl."

Northern Virginia is by no means alone when it comes to sprawl. Cities from Atlanta to Los Angeles, Austin to Chicago, and everyplace in between have seen the uncontrolled expansion of their urban areas. But the ground here is unique in our country's history. And symbolic, perhaps, of the country at large.

I'm setting out on a driving tour of Civil War battlefields that will take me progressively east across Northern Virginia toward Washington, DC, and this valley is where I begin. Tucked between the Appalachian Mountains in the west and the Blue Ridge Mountains in the east, the Shenandoah Valley runs north–south through Virginia. The worn-down ridges of its old mountains slope with deciduous trees that in autumn turn to brick, orange, and gold. With evidence of human presence going back at least fifteen thousand years, the valley has long been a fertile agricultural area, and during the Civil War was the "breadbasket of the Confederacy." When the war's tide turned decisively for the North, Union general Philip Sheridan advanced through the valley, systematically burning everything in his path.

"I had a glimpse of the past," says Kehoe, a tall, bespectacled man with a salt-and-pepper beard. He's known the Cedar Creek Battlefield all his life, growing up in midcentury, when local farmers still used horses to plow their fields. Sometimes these farmers would show him an arrowhead they'd turned up, and would tell him stories of their grandparents in the Civil War. "That touch with the past is gone," he says while showing me the museum at Hupp's Hill. As a boy, he walked this ground with his grandfather, himself a cavalryman from the First World War who one day kicked up a sword from among leaves in the woods. Thus began a lifelong obsession with artifacts, one with which he's had a lot of success. For Civil War diggers, finding a Confederate belt buckle is, Kehoe says, "like the twelve-point buck." Here in the museum, in one display after another, Kehoe shares his findings: more than two hundred belt buckles, five hundred Confederate buttons, cannonballs, thousands of bullets, "everything," he says, "that you could find that soldiers lost." From the start of the war in 1861 to its conclusion in 1865, infantry, cavalry, and artillery from both armies moved up and down the valley. And apparently they lost a lot of stuff.

When the construction of Interstate 81 came through the valley in the early 1960s, Kehoe followed the bulldozed path. "We would go over on Sundays when the construction workers shut down for the weekend, and it was just bullets laying everywhere, and buttons." He pauses, then chuckles. "I guess I was a preservationist even then, because my brothers and I, when the surveys were coming through and we knew it was going through the heart of the battlefield, we took our wagon and picked up the survey stakes and used them for kindling."

It's probably ironic that I travel to visit Kehoe on I-81, and when I leave I will head east on I-66. Kehoe calls I-81 the "biggest impact on the Shenandoah Valley since the Civil War," a statement that could no doubt be applied to other interstates around the country. Ostensibly designed to facilitate the movement of American troops in case of war, the Interstate Highway System was, historians argue, mostly a way to encourage development of previously remote areas. On the hour-long drive north from Harrisonburg, up through the valley, my route passes a steady stretch of warehouses, big-box stores, and truck stops, each exit hived with gas stations and fast food.

"It's a blight to our landscape," he says as we take the short drive from museum to battlefield in his pickup. "I'm not against development, but the sprawl aspect of it just breaks my heart." He points to a fifteen-acre site where a warehouse holds toilet paper imported from China. "It's a waste of our land, and I think we'll pay for it sometime down the road, when agricultural land is going to be more important than industrial land. And we're already paying for it because of the time we waste commuting. The quality of our life is not as good as it used to be." By way of example, Kehoe tells me that his wife commutes an hour each way back and forth toward DC, which is not an uncommon commute in the United States. That's ten hours a week, and at fifty weeks a year, the equivalent of spending seven weeks each year trapped in a car.

Recent black-and-white maps of the United States, with every other feature removed—no lakes, rivers, mountains, cities, towns, parks—reveal how thoroughly our roads shape our lives. Our cities are black knots, much of the eastern part of the country is a web so dense the individual lines blend together, and only a few small blank areas remain—roadless areas—mostly in the West. Strangely revealing and even beautiful, these road maps show that even "out west" the country is covered. In fact, in the lower forty-eight now it has become quite difficult to get away from roads—the farthest you can get is some twenty miles from the nearest road, and that is definitely the exception. Often, even if you can't see the nearest road, you can hear it. These maps reflect a shocking truth — roads are just about everywhere.

Even so, in terms of actual area, roads themselves don't cover that much ground. According to the Federal Highway Administration, there are now four million miles of roads in the United States. (A mere 225,000 miles of the National Highway System carries most of the traffic in the nation, and three million miles of roads are rural, more than half of those unpaved.) That's a lot of road, but in the context of the nation's great size, not that much. In fact, the American Road & Transportation Builders Association answers the question "Are we 'paving over America'?" by pointing out that even with shoulders, driveways, and parking lots included, "roads" only cover about 1 percent of the nation's land area. That may not sound so bad (as they are quick to emphasize), but numbers tell only part of the story.

"Roads scare the hell out of ecologists," says William Laurance of Australia's James Cook University. "You can't be in my line of business and not be struck by their transformative power." That transformative power comes from the fact that when we slice and dice the ground with our roads, we weaken the ecological fabric on which life depends. Ecologists call this fragmentation, and its costs are enormous.

A recent study led by Nick Haddad of North Carolina State University found that "fragmented habitats lose an average of half of their plant and animal species within twenty years, and that some continue to lose species for thirty years or more. In all of the cases examined, the worst losses occurred in the smallest habitat patches and closest to the habitat edge." Alarmingly, Haddad and his co-researchers found that "more than seventy percent of the world's forest now lies within one kilometer of such an edge."

Not all habitat is equal, something that's not necessarily obvious when we're driving through fields or forests, glancing at passing trees. But just because the woods may come close to the road doesn't mean that those woods are being much used. Scientists have found that wildlife will tend to stay away from those edges and move to the interior of the forest. Which makes sense; with its roaring machines, the edge of the forest means danger to animals. With our road building we have created more and more of those habitat edges, leaving less and less of the "interior habitat" that wildlife most prize. Whether it's in northern Virginia or northern Brazil (or any direction anywhere else), fragmentation reduces the actual amount of habitat animals can use.

With the world's human population projected to everywhere continue its growth, the problem promises to become worse. New research from the University of Cambridge suggests that "more than 25 million kilometres of new roads will be built worldwide by 2050," and that "many of these roads will slice into Earth's last wildernesses, where they bring an influx of destructive loggers, hunters and illegal miners." One of the study's lead researchers, Christine O'Connell of the University of Minnesota, says, "Our study also shows that in large parts of the world, such as the Amazon, Southeast Asia, and Madagascar, the environmental costs of road expansion are massive." The problem, O'Connell and her fellow researchers understand, is that with more than nine billion people by 2050, the world will need more roads. But they hope that by better understanding the devastating effects of reckless road building they can encourage stronger planning before those new roads are built. "So much road expansion today is unplanned and chaotic, and we badly need a more proactive approach," says Laurance, one that will guide the development of new roads to avoid unnecessary environmental costs. "It's vital because we're facing the most explosive era of road expansion in human history."

Scientists tell us that all around the world, biodiversity is in decline. A recent study published in *Nature* begins, "Biodiversity faces growing pressures from human actions, including habitat conversion and degradation, habitat fragmentation, climate change, harvesting and pollution." In the United States, "biodiversity has decreased...by nearly a third." Tropical protected areas "are suffering serious biodiversity declines." But what does all this really mean? Sometimes the loss of biodiversity is obvious—there are no grizzlies in Central Park—but usually not. In fact, as one ecologist told me, "99.99 percent of biodiversity is stuff we never see."

Some biodiversity we never see because it's simply too small. But we also do not see biodiversity because we are part of it; we live immersed. I used to hear "biodiversity" and think it had something vaguely to do with rainforests. It didn't occur to me that it was everywhere, that it meant the life of the whole planet. Nobody knows who discovered water, goes the old saying, but it probably wasn't a fish. As Wake Forest University biologist Miles Silman told me, "Diversity isn't something that exists in the world; diversity makes the world. The things that regulate the populations, the things that keep the pests in check, the things that structure the amounts of the different kinds of fishes in the seas— they're all based on biodiversity and the ecological interactions between the organisms. So if you lose it, you're going to reshape the world."

Nothing speeds along that reshaping like roads. Certainly in the United States, without roads—and the motorized transport they host—our development would be much more compact. Instead, roads have allowed us to spread across the country. Paved roads bring paved grounds, ensuring that we have less contact with the natural ground than any civilization before us.

"To build a road is so much simpler than to think of what the country really needs," wrote the American ecologist Aldo Leopold in the 1940s. Best known today for his book *A Sand County Almanac*, Leopold had plenty to say about our relationship with the planet. Something of a hero for me, he will show up on these pages a few more times. In his essay "Marshland Elegy," he mourned the road building so destructive to cranes and other creatures. And how, to the Western mind, "a roadless marsh is seemingly...worthless."

In fact, experts say, roadless areas are invaluable. Many species need large areas of unbroken habitat to survive. This is especially true of megafauna like wolves and grizzlies, what ecologists call "indicator species," those species at the top of the food chain whose presence indicates an ecosystem's health and whose absence indicates illness.

But our habit of road building is difficult to break. When in 1999 President Clinton attempted to preserve what roadless areas remained in the United States, opposition was fierce. How dare he "lock up" areas that might otherwise be developed? In fact, the "roadless rule," as it came to be known, proposed to protect only areas on federal land, mostly in national forests and national parks, and represented only a small percentage of the nation's total land area. Despite opposition, the roadless rule became law, and while far from perfect, it made a difference. Just as the negative effects of roads spread beyond the pavement, so the positive effects of roadless areas spread beyond their relatively small square mileage. In many cases these roadless areas represent the world's last intact ecosystems, the last ecosystems able to support the large predators that once roamed the earth. They are also, oftentimes, the last best reserves of the world's biodiversity.

Antietam, Shiloh, Chancellorsville, Spotsylvania—Cedar Creek does not rank among the well-known battles of the American Civil War. But the fight here was significant in the last part of the conflict: it's the place where the Confederate claim to the Shenandoah Valley was finally broken, and where more than nine thousand Americans fell. Mike Kehoe and I pull off US Route 11 near where the Eighth Vermont Regiment lost 110 of its 164 men while slowing a Confederate advance. Kehoe tells me the state of "*Ver*mont" lost more soldiers at Cedar Creek than at any other battle, and says, "They were saving the Union. Those guys are why we're here. If it weren't for them it would be two different nations, and who knows what."

What fascinates me about my host is that he has known a piece of ground intimately all his life. The grounds here are little changed from the scene of the 1864 battle, and Kehoe has roamed them since he was a boy. Once, a stick emerging from mud turned out to be a rifle; another time, playing in the creek, he caught his foot in a cavalry sword handle. Over the course of his life he's learned from archeologists and historians, and from reading the diaries of those who were here, including two young soldiers from North Carolina who, he says while pointing west, died on the creek bank right over there. Unfolding a cloth on the pickup's hood, he says, "This is a good time to show you what I found this week." He lays out a broken bayonet, several heavy bullets, and a Union button flecked with gold. "There's still a lot in the ground," he says.

In a ravine to the west, maybe a hundred yards away, stands an eighteenth-century farmhouse that during the battle was a Union hospital. "When we get down there," Kehoe says, "I'm going to show you some of the features in the ground. They would have buried the soldiers that died there away from the house, and I think I know where it is." After a short hike over a slope that hides the house from view, Kehoe points to several rectangular depressions just visible in the grass. "I think this is where they buried the soldiers," he says. After the battle the bodies would have been transferred to state or national cemeteries, but, he explains, "there are probably still bones here, finger bones, little bones. A lot of times they only transferred the skull and the bigger bones." And here's the thing: until Mike Kehoe found this spot, no one else knew it was here. No grave markers, no headstones, no black iron fence marks the spot. He just knew that wounded soldiers would have died in the hospital house and been buried somewhere nearby. "And if there are sites like this here," he says, "then they're all over Virginia."

Increasingly, though, sites like these—whether known or unknown—are themselves being entombed. Kehoe tells me of a site near Richmond where he once dug up four Confederate rifles—"but it's a housing development now." Even at Cedar Creek, east across Route 11, an enormous Thermo Fisher plant blocks the view. To the west, a dozen houses have been added in the past decade or two, and our short trip back to Hupp's Hill takes us through the usual strip of convenience stores and gas stations so common in American cities of every size. Each of these buildings covers ground where Civil War soldiers once camped and fought and died. As Kehoe drops me back at my car and wishes me well on the rest of my drive, he says, "I would say more than half the sites I know of—and I've been digging since the nineteen sixties—that we used to go to are gone, eliminated. And," he adds, tipping his hat, "more will be gone soon."

Why? A growing population, a style of development made possible by roads, and a faster paving of natural ground than at any time in history. In fact, a third of all the natural ground ever developed in the lower forty-eight states has been developed in the past thirty years. In the United States from 1980 to 2010, more than forty million acres of natural ground—forests, wetlands, prime farmland—were developed, much of it paved. That's an area equal in size to the six New England states plus New York and Pennsylvania. But the two key characteristics of sprawl—size and speed—are evident all around the country. In the Chesapeake Bay region, more land will be developed between 1995 and 2020 than in the previous 350 years. The Phoenix metropolitan area, already equivalent in size to the state of Delaware, continues to steadily digest its surrounding desert at more than an acre an hour. And of Atlanta, says one expert, "it is altogether probable that in terms of land area Atlanta is the fastest-growing human settlement in history."

Where do we go from here? At the century's turn in 1900, the United States had a population of 76 million people. A hundred years later, we had 282 million. We already have 325 million in 2016, and most projections have us rising to at least 400 million by 2050 — 75 million more people who need to live somewhere, and get to somewhere else.

About fifty miles east of Cedar Creek on I-66 lies Manassas National Battlefield Park, the site of the first major battle of the Civil War. The Battle of First Manassas (or, to the Union army, First Bull Run) took place July 21, 1861. We remember it as the battle that gave Confederate general Thomas Jackson his nickname "Stonewall." Famously, it was also a battle in which young soldiers on each side were stunned by the realities of death and destruction, and the Union retreat dissolved into chaos amid the carriages of congressmen and socialites coming through nearby Centreville to witness "the grand spectacle of war."

Today I am most interested in how one of the country's most famous battlegrounds, already crisscrossed by US Route 29 and Virginia Route 234, has twice been nearly overrun with sprawl. Exiting I-66 takes me immediately into a commercial area featuring an Old Country Buffet, a Cracker Barrel, a Holiday Inn Express. I join a long line of cars on 234, passing a community college before turning in to the battlefield park. A curving lane brings me to a visitors' center, where on the wall inside a soldier from Georgia expresses the shock of actual combat: "My first thought was, 'This is unfair; somebody is to blame for getting us all killed. I didn't come out here to fight this way; I wish the earth would crack open and let me drop in.'"

A back door releases me onto the battlefield. Essentially unchanged since 1861, it's easily walkable, as most of the fighting was contained on a grassy plateau called Henry House Hill. Along the walking loop, small black signs tell me that one soldier or another "was killed here." At the southern edge of the battlefield, the site where the First Minnesota fought, forty-nine were killed and many more wounded—the highest casualties of any Northern regiment. By dusk, the Union army had retreated and nine hundred young men were dead. "Up on the bluff we saw the first dead Yankee," wrote one soldier from Virginia. "The pale face turned towards us, upon which we looked with feelings mingled with awe and dread."

The visitors' center parking lot covers the ground where some of this dying took place. Standing with one foot on asphalt and the other on ground never paved, the bare ground feels alive, filled with stories that you could dig into literally and figuratively. The paved ground feels dead.

In 1988, developers purchased more than five hundred acres of private land next to the battlefield and proposed a 1.2-million-square-foot shopping mall, a threat that was averted only with an emergency appropriation of $120 million from Congress. Five years later, the Walt Disney Company announced plans for a major "historic theme park" that would include three hundred RV campsites, nearly two million square feet of retail and office space, and twenty-three hundred houses. Although many politicians eagerly welcomed the proposal, in the end a coalition of groups and famous individuals persuaded Disney to withdraw its plan. Good news, as one historian noted, but "the slow advance of development continues. More than likely, there will be no national outcry if a filling station is put here or a small residential subdivision there, but the end effect for the park will be the same: the replacement of the last bits of open fields with modern life; the steady stream of more and more cars on the fragile two-lane roads."

On the day of the battle, arriving soldiers from both sides

followed the sound of the fight. But here the highway is loudest now. When I leave, turning east on 29, it's rush hour, the two-lane highway mobbed, and just beyond the battlefield's boundaries begins development that runs without interruption from Prince William County to Fairfax County to where Lincoln sits in his memorial, mulling over what's become of the country.

If any county in the United States is the epitome of sprawl, it's Fairfax County, Virginia. Immediately west of the Capital District, Fairfax has seen its population grow from 450,000 in 1970 to more than 1.1 million in 2015. The estimates for 2040 are of a population nearing 1,450,000—a million more people on the same amount of land in my lifetime. And if any one area of Fairfax County epitomizes the development so prevalent in Northern Virginia, it's Tysons Corner.

During the Civil War it was a bucolic crossroads known as Peach Grove. It was farmers' fields as recently as the mid-1960s. But Tysons Corner had by 2010 been variously described as "a city for cars," "a shrine to anonymous suburban sprawl," and "the blob that ate Northern Virginia." Though fewer than twenty-five thousand people actually live in Tysons (the city will soon drop the quaint "Corner" from its name), every weekday more than one hundred thousand people visit—and they visit by car.

I decide I will join them. From Manassas, I set my GPS to "Tysons Corner" and follow its lead—straight to a shopping mall parking lot. I shouldn't be surprised. Key features of the city are its two enormous malls and more parking spaces than you have ever seen—some 167,000 at last official count. Parked between a massive Nordstrom and a giant L.L.Bean, I strain to see where I might stand on anything other than pavement. If the ground at our feet says much about who we are, here we are well-to-do shoppers on our way to spring pastels. As a thunderstorm moves through, its rain straining to find ground to seep into, I try to imagine farmland. But here that past has been erased, as in parking lots all over the country.

In fact, while the official estimate is more than one hundred million parking spaces in the United States today, some studies suggest closer to two billion. This would mean eight parking spots for every car and truck in the country. Eran Ben-Joseph of MIT reports that "in some US cities, parking lots cover more than a third of the land area, becoming the single most salient landscape feature of our built environment." Much of this parking acreage exists because of zoning rules requiring specific ratios between parking spots and housing units, or parking spots and square feet of retail space. In one example, Stockton, California, required of developers one parking spot for every 250 square feet of retail space — so a mall of 750,000 square feet resulted in more than 3,000 spots. And while malls come in a range of sizes, dozens of malls in the United States take up more than a million square feet. Another reason for so many parking spots? Planners, developers, and retailers often rely on the idea that there should be enough parking to accommodate the busiest shopping day of the year — though no one can say exactly how many spaces that would be, and everyone knows of a parking lot that sits mostly empty most of the year.

It's worth emphasizing that Tysons, Virginia, is just one example of the sprawling car-centered growth taking place worldwide. The world already has more than five hundred million automobiles, but more than four-fifths of those are in the industrialized countries. Now, reports Lester Brown, "in developing countries, where automobile fleets are still small and where cropland is in short supply, the paving is just getting underway." China alone could eventually pave upwards of thirty million acres of land to support the more than five hundred million *new* automobiles it would have if car ownership rates there ever match those in the United States. Part of this will be parking, of course—in 2015, China built nearly half the new mall space in the world.

Meanwhile, still parked in Tysons, I type "Ox Hill Battlefield Park" into my GPS. The only major Civil War fight to take place in Fairfax County, the Battle of Ox Hill (to the Union army, the Battle of Chantilly) saw more than fifteen hundred soldiers killed or wounded during a thunderstorm loud enough to drown the sounds of battle. Only twenty miles from the Capitol building, here the Confederate general Robert E. Lee attempted to outflank a retreating Union army. Union divisions under the command of General Isaac Stevens and Major General Philip Kearny engaged the Confederates, and in doing so, both generals were killed. I know the battlefield can't be far from here, but MapQuest says, "Well this is no fun...we can't find that exact place."

Possibly that's because nearly the entire battlefield has been paved. (Despite this, having found the park's web page, I'm warned that "Parking is extremely limited.") Eleven miles later, I reach what a Dickensian Ghost of Civil War Battlefields Future might bring me here to show: the once-sprawling battlefield shrunk to barely a block and encircled by apartment buildings and office parks. Of the original five hundred acres on which the battle was fought, the park protects about five. Hark, who goes there through the tree trunks and leaves? A half-dozen boxy beige four-story condos, silent in their advance.

"The Civil War's battlefields are precious to us," wrote *National Geographic* in 2005, "not just because of blood and heroism, but because they let us glimpse a vanishing, agrarian America—the very landscape in which our national identity formed." Ox Hill/Chantilly is often cited as what can and will happen if people don't act. The good news? What happened here sparked the creation of the Civil War Trust, an organization that since the mid-1980s has helped preserve more than forty thousand acres of Civil War battlegrounds threatened by development. Still, more than ten thousand privately held acres are lost every year. And with about two hundred thousand acres of unprotected Civil War battlegrounds left, in about twenty years the battle will be over.

What else will the next twenty years bring? Assuming that our population grows as expected, it's likely that development will continue across what remains of the planet's natural grounds. That's true in the United States and it's true worldwide. One recent study found that in the western United States a football field's worth of natural land disappears every two and a half minutes, another that "urbanization, agriculture, and energy could gobble up 20 percent of the world's remaining natural land by 2050."

I keep thinking of beloved places. The Minnesota city of my birth. The chopped-down childhood forests my students tell me stories of every year. I keep thinking of this northern Virginia farmland, and of people like Mike Kehoe, who has lived here all his life. I keep wondering, As a society, is this what we want?

The wet rush of tires on asphalt, the groans of packed planes gaining height, leftover storm clouds blotched by streetlights and heat lightning. The old songs of frogs, the older smell of rain. On this ground, in driving rain, young Americans desperately killed each other, fighting hand to hand on behalf of their brothers, their families at home, for what they believed right. Here in the dark, the thunderstorm having passed through, I step off pavement with bare feet onto slick grass, trying to make sense of where I am.

Gettysburg

At what point shall we expect the approach of danger?...Shall we expect some trans-Atlantic military giant to step the ocean and crush us at a blow? Never! All the armies of Europe, Asia, and Africa combined...could not by force take a drink from the Ohio or make a track on the Blue Ridge in the trial of a thousand years....If destruction be our lot we must ourselves be its author and finisher. As a nation of free men we must live through all time, or die by suicide.

— ABRAHAM LINCOLN, LYCEUM ADDRESS (1838)

I think of them now, running toward me in their baggy wool blues, fingers gripping long rifles, shouting maybe, gritting teeth. I bet they trip, some, the sloping ground uneven, stone-poked, the smoke swirling, blocking views. They don't think of our home, of Minnesota, they don't think of the future or lovers or even death. They are just running. Lead shocks and spins friends, the *splat-thud* of shot minie ball meeting flesh, and they stumble steps away, palms scraped, stained, breaking falls. But they keep on, all 260-some charging this way. And about here, near where I kneel, they meet the enemy—same language, same fears and loves, same country in the end—and in twenty minutes it's over, the Union saved, more than 80 percent of the First Minnesota killed or wounded, spread across this ground that at dusk soaks in their blood.

Months later, close by, the president will acknowledge their sacrifice, what they have done—they and countless others—saying that it's up to us to remember, that they have done their part, making this hallowed ground. And decades later, dusk again, I'm here alone with them. Lying in long summer grass, inhaling the smell of soil close, wondering if any that fell knew the same. The fight done, was it quiet like this? Were frogs singing just past spring, killdeer calling in what seems delight, swallows carving a darkening sky? Were fireflies and lightning bugs floating skyward by the dozens, then hundreds, as many as the dead and wounded that will lie in surrounding fields tonight?

Ask an American to name "hallowed ground," and they will say "Gettysburg." For us, the words are nearly synonymous. Here, in early July 1863, a battle raged that was the largest and bloodiest ever on United States soil. In three days of fighting at least eight thousand men died on this battlefield, and many more died soon afterward of their wounds. The fighting was incredibly intense, the battle considered the turning point of our Civil War. While others helped to shape the idea of this place—the president's invitation to the event read "These Grounds will [be] Consecrated and set apart to this Sacred purpose"—it is Lincoln's short speech we remember, and his argument that "In a larger sense, we can not dedicate—we can not consecrate—we can not hallow—this ground. The brave men, living and dead, who struggled here, have consecrated it far above our power to add or detract." Some of those brave men came from my home state of Minnesota, and it is to their place on the battlefield that I have come. I want to know what, in the end, "hallowed" means.

To our modern ears, the word sounds like something

from the nineteenth century, from fading photos in black and white. And at first look this hallowed ground does not seem much different than any other. There is no magic aura floating above the blades of grass, no special tinge of color to the soil. The small flowers are the small flowers I have seen elsewhere—buttercups, daisies—and the birds do not fly surrounded by a golden glow. That's one of the things I notice immediately about hallowed ground—it wouldn't feel hallowed if no one had told you it was. Or maybe I'm wrong.

There are so many famous scenes from this battle, and at Gettysburg, everything is about preserving those sites. In addition to simply protecting the grounds from development, the National Park Service is actively engaged in returning the grounds to what they looked like during the battle. In one location it might cut down mature trees to re-create a farmer's field, and in another it might plant trees where they would have been. In both preserving and re-creating, it has plenty of challenges. During the battle there were thirty-eight fruit orchards, and today there is only one, the famous Peach Orchard, where heavy fighting raged. There is the odd modern spectacle, on this hallowed battlefield, of sharpshooters hired to cull a deer herd burgeoning far beyond its Civil War–era numbers. And there are the many dozens of monuments honoring the various states, most at places where units from that state had their moment of glory. A handful of these monuments draw bigger crowds than others—perhaps made famous in a movie or novel, or honoring soldiers from nearby states. For example, here in Pennsylvania the massive Pennsylvania monument is particularly popular. But next door, at the base of the monument honoring the soldiers who came from farthest away, I am alone.

This monument has a white base topped by the bronze

figure of a Union soldier running with rifle and bayonet in the direction the First Minnesota made their charge. It was July 2, the second day of the battle, and just before dusk. The Minnesotans were guarding a battery of cannons when Alabamans began to advance. If the Confederates were able to break through the Union line here, it would split the Union army in two, and quite possibly turn the battle and thus the war.

The Union's commanding officer in the area, Winfield Scott Hancock, rode to where the soldiers from Minnesota lay listening to the rebel minie balls "hissing over us, cutting the weeds and bushes, plumping into the ground and spatting against the stone," and said, "My God, are these all the men we have here?" It was. No matter. Without hesitation he ordered the First Minnesota to charge Confederate forces four times their number. "Every man realized in an instant what that order meant—death or wounds to us all," wrote one of the soldiers, "the sacrifice of the regiment to gain a few minutes' time and save the position, and probably the battlefield." And the next thing the Alabama regiments saw coming toward them were 262 Minnesotans, running forward, bayonets fixed.

It took the Union soldiers about ninety seconds to cover the three hundred yards between themselves and the Confederates, most of it a harvested wheatfield's stubble. Ninety seconds of double-timing down the slight slope toward the trees and bushes where the soldiers from Alabama were shocked to see "screaming men charging out of dense smoke." But not so shocked that they couldn't unleash waves of rifle and cannon fire that tore through the Minnesota ranks. "Great heavens how fast our men fell," wrote one soldier. "It seemed as if every step was over some fallen comrade. Yet no man wavers: every gap is closed-up." The "double-quick" march with which the charge had begun quickly turned to as-fast-as-they-could-go, for it was the "only hope that any of us would pass through that storm of lead and strike the enemy." Captain Muller was shot through the head and dropped dead. First Lieutenant DeGray fell with a four-inch rip in his skull from a bullet. Captain Perium had a minie ball "punch through the right side of his nose and exit behind his left ear." The Minnesotans charged to within feet of the Alabamans, then let fly bullets of their own, and the two armies met in hand-to-hand combat in a dry ravine called Plum Run.

Renowned historian James McPherson tells me the noise is perhaps the hardest thing for a visitor to imagine. When he visits Plum Run, "What I imagine is chaos, confusion, yelling. It's pretty hard to put yourself into the cacophony." The noise, and the smoke. As the Minnesotans' commanding officer William Colville wrote later, "My glance took in the slope on my left. I saw numbers of our men lying upon it as they had fallen. Then came a shock like a sledgehammer on my back bone between the shoulders. It turned me partly around and made me 'see stars.'" He had been hit by a minie ball that nicked his spine and lodged under a shoulder blade. Putting his foot down to steady himself, he was promptly hit in the ankle with yet another ball, then pitched forward and hit the ground. Nearby, a shell burst above Private Isaac Taylor, taking off the back of his head and nearly slicing him in half.

At dusk, the fighting subsided. The Confederate commanders realized that their forward momentum had been stopped. When the Minnesotans rallied back around the unit's colors, they found they had only twenty-five men.

"Every man, without exception, had his clothing riddled—some of them all to rags," wrote the regimental quartermaster. With both sides pulled back, hogs unpenned from nearby farms entered the battlefield, devouring some of the dead. In a letter home a soldier wrote, "The ground was strewed with dead and dying, whose groans and prayers and cries for help and water rent the air. The sun had gone down and in the darkness we hurried, stumbled over the field in search of our fallen companions." Another soldier slept, he wrote, "with dead men and horses lying all around me."

The casualty rate suffered by the First Minnesota that day—82 percent killed and wounded—was the highest of any unit in the Union army in any battle of the Civil War. But in general and for all, Gettysburg was a battle of carnage. McPherson reports that after the battle ended, "burial details hastily interred more than three thousand dead Union soldiers and many of the almost four thousand dead Confederates. Five thousand dead horses were doused with coal oil and burned." In locations that saw particularly heavy fighting, such as the famous Wheatfield, soldiers reported so many dead and wounded that they could have walked over the battlefield without ever touching ground. One wrote,

> Corpses strewed the ground at every step. Arms, legs, heads, and parts of dismembered bodies were scattered all about, and sticking among the rocks, and against the trunks of trees, hair, brains, entrails, and shreds of human flesh still hung, a disgusting, sickening, heart-rending spectacle to our young minds. One man has as many as twenty canister or case shots through different parts of his body, though none through a vital organ, and he was still gasping and twitching with a slight

motion of the muscles, and vibrations of the pulse, although utterly unconscious of approaching death.

"For months," McPherson writes, "the stench of hospitals, and of corpses unburied or buried in shallow graves, hung over the town and countryside."

Many of the most famous photographs from the Civil War are from Gettysburg, and of these most are of Confederate soldiers, their bodies bloated and faces contorted in death. As Drew Gilpin Faust has written in *This Republic of Suffering*, the sheer number of casualties overwhelmed the two armies' burial capabilities. Because the Union controlled the battleground, its dead received better attention than those from the Confederate army, many of whom were simply dumped into mass graves "like dead chickens," or left to rot. This is why most of the famous photographs are of dead Confederate soldiers showing the first stages of decay: rigor mortis, and bloat as microbes grow and form gases in the body (the next stages were yet to come: loss of mass due to insect eating and release of liquids, advanced decay where there is little left of the body, and finally skeletonization, where no flesh remains).

Isaac Taylor's brother Henry found his sibling the next day, and with two comrades dug his grave: "We laid him down with all his clothes on, as he fell, and spread a shelter tent over him." Battlefield conditions, as well as a lack of time and supplies, dictated shortcuts for the care of the bodies, even for the best of friends or brothers. Grave markers were made from scrap wood and marked with pencil. Not surprisingly, twenty-two bodies now lie in the Minnesota plot of the Gettysburg National Cemetery with nothing but "Unknown. Regt. 1" on their gravestones.

It was nearly eight years before organizations from the former Confederate states returned to disinter their dead, and by then identification was difficult and finding all bodies impossible. As recently as 1997 a skeleton—most likely of a Confederate soldier—emerged from the soil. "I'm sure he's not the only one," McPherson tells me. "They obviously didn't find everybody, and I'm sure there are a lot more. Which makes the whole battlefield something of a cemetery."

Remarkably, despite the momentous events that took place on these grounds, it took an 1896 Supreme Court decision to guarantee the battlefield's protection. In the case, the government argued, "the ground whereon great conflicts have taken place, especially those where great interests or principles were at stake, becomes at once of so much public interest that its preservation is essentially a matter of public concern." The court agreed, and declared that protecting the field of battle was "so closely connected with the welfare of the republic itself as to be within the powers granted Congress by the Constitution for the purpose of protecting and preserving the whole country."

At 6:45 on a summer's evening, I start down the slope where the First Minnesota charged. Rather than a field of harvested wheat, I walk over clover-crowded ankle-high grass. Small yellow butterflies. Killdeer, swallows. The highway in the distance steadily hums with cars. Mountains on the horizon, open fields ahead and behind, barns in the distance. If I were a Minnesota soldier I'd be double-timing it downhill toward the trees. I'd be jogging toward the ground where I might die. This evening, I'm alone. I wonder if there's anything left of the Minnesota out here, in the ground. My friend Laura, who grew up in northwest Virginia, says when she buried her beloved dog in her childhood backyard she found a Confederate belt buckle in the dirt.

When I reach the ravine, where the worst fighting took place, the hand-to-hand, the bayonet-stuck bellies and the rifle-crushed skulls, I step on an exposed rock and think, *This was here. I am on the same ground where those soldiers from Minnesota fought and fell, at the same time of day.* I drop to my knees in thigh-high grass, then fall forward to rest my face against the ground. The smell of grass, clear beads of water on long green leaves, the sight of a ladybug's burnt orange. When I rise, the impression of my body, my knees, stays. Ground soft from rain, wind in my hair. I'm walking where they did. Impossible to know exactly what happened where on this blood-soaked ground. Bumpy, grassy ground, brushing bare calves. Decaying logs, clovers, beds of tan-brown fallen grass. Buttercups, daisies, brown-eyed Susans. The stars of flowers, the stars above.

I wonder, How different is this Gettysburg ground from what it was? How has this ground changed over the years?

John Commito came to Gettysburg College in 1993, bringing a thick Boston accent and an expertise in ecology. Every September since, he has started his Principles of Ecology course on the battlefield, and one big theme is the changing ground. "That's a very important lesson for students to learn. Because of human factors but also because of nature running its course through time, the way it looks today is not only different from the way it looked in the eighteen sixties, it's different from the way it looked ten years ago. It is a living landscape. It's not paved concrete."

That said, when I ask Commito how similar the actual ground of the battlefield today is to what it was during the battle, he tells me that though the specific soil may have changed, the level of soil is generally unchanged. And that the larger the item—think of a gradient from soil to pebbles to cobbles to boulders—the more likely it is to remain in place. "So," he says, "certainly there are witness stones on the battlefield—a stone that was present during the battle— even though the small soil particles may have been eroded away and replaced by new soil particles."

I ask whether for him, knowing it's the same ground makes walking on the battlefield feel different than walking elsewhere. "For me it definitely does, and for the students too. One of our study sites used to be by the original line of Pickett's Charge. Everyone knows about Pickett's Charge, and everyone knows about the slaughter that took place. And that's right where we get out, right by the Virginia monument. Then we walk into the woods and we do our work. That definitely freaks out some kids."

And does the idea of "hallowed ground" have anything to do with the actual ground, or is it just what happened there? "Well, a lot of people died there," Commito says. He tells me that he teaches a course in which he takes the class to a graveyard, and that he will often have students start to cry, "usually when they see a little baby's tombstone." And that when he takes them out on the battlefield and they know that thousands of people were killed and wounded right where they're standing, they feel close to what they felt in the cemetery. But he says that students also feel this way in the forest, when "they see the big giant dead trees, and it impresses them when they see a tree that is dying or has died, and is lying down on the ground, and it's going to enrich the soil with organic material and nitrogen and phosphorus. And gee, that thing was around for a hundred and fifty years, and then it fell to the ground—nothing is permanent. Maybe they get some of the same feeling when they know that all of these people who were their age were killed right there where we are going to be collecting data—literally under our feet."

A colleague of Commito's at Gettysburg College, Kent Gramm is the author of *Gettysburg: This Hallowed Ground* and *November: Lincoln's Elegy at Gettysburg*, a book that was nominated for a Pulitzer Prize. When I ask him about hallowed ground, Gramm says that over the years he has talked to many visitors on the battlefield who come with a sense of pilgrimage.

What they're looking for, he says is "not necessarily something in a patriotic sense," but something larger, something like "the meaning of life, or the presence of the divine." And, he says, he thinks people have been drawn to the location for centuries. As he explains, Gettysburg was sacred to Native Americans long before the battle, and they used it as a meeting place because of the strange geological formations on the southern end of the field. "Or maybe," he says, "it's because a place of great stress and impact still retains something, as Joshua Chamberlain said."

It turns out that Chamberlain returned many times to the scene of the Twentieth Maine's battle on Little Round Top, and during one such visit said,

In great deeds, something abides. On great fields, something stays. Forms change and pass; bodies disappear; but spirits linger, to consecrate ground for the vision-place of souls...generations that know us not and that we know not of, heart-drawn to see where and by whom great things were suffered and done for them, shall come to this deathless field, to ponder and dream; and lo! The shadow of a mighty presence shall wrap them in its bosom, and the power of the vision pass into their souls.

"You could believe that or not, I suppose," Gramm says. "A lot of people go to Gettysburg and find that there's just something there that gets you. It's moving. But what if they didn't know? Is it something we bring? Is it something in our minds?"

I think so. It's how we think of certain grounds that dictates how we use them. Why do we consider certain ground sacred or hallowed, for example, and not other ground? Why do we keep from paving or developing some areas but with others plow ahead?

Gramm believes it's because "we have lost a sense of the sacredness of human life. When we visit a place like Gettysburg, we're really confronted with the sacredness of human life when we see how fragile it is. It reminds us because it's so compressed a place where so many gave their lives. But in a way, we should have this consciousness everywhere. When we desecrate the earth, and desecrate our own place of living—and where others have lived, and worked, and struggled, and played, and died—I think that shows a nonvaluing of human life that a battlefield briefly reminds us of."

Could the counterargument be, Well, that's just progress? I ask.

"Yes, progress. It's almost like doublespeak, isn't it? It may be progress for some people who get what they want, while the rest of us have something destroyed."

For Gramm, there's definitely a presence on Gettysburg's hallowed ground. "What I feel," he says, "is not necessarily war related. I feel a divine presence somehow. I don't know whether to explain it because of the violence and death, and the sacrifice, and the nobility and courage, or because of any spiritual presence that's somehow related to that. I don't know why or how it would be related to that."

He ponders his own question, and then comes back to Lincoln. Maybe it's the idea that many people were there for a sense of something beyond, and higher than, themselves, he says. And in that case, it can apply to Southerners as well. "Even though it was not the cause that Lincoln is saying was sacred," he reflects, "perhaps the sacrifice for something more than oneself—whether mistakenly or not—carries with it some kind of holiness. If you are willing to fight and sacrifice for others, that means you're fighting for something higher, and at Gettysburg I think that higher purpose connects with equality. All the major religions speak of transcending that illusory self that we are so preoccupied with and that keeps us in the grip of sin. When we can somehow transcend that false self and connect to, if we use Paul Tillich's term, 'the ground of our being,' there's something sacred about that." Later in my travels I would hear more about Tillich, the German theologian who, in perhaps his most famous essay, replaced the term "God" with the phrase "the ground of our being."

"In a political and historical sense," says Gramm, "Gettysburg and the Gettysburg Address are the ground of our being. Because that's who we are. That's the only way that being American makes sense and differentiates us from anyone else."

FARMED AND WILD

The grounds that feed us, body and soul.

Bishopstone

Eating with the fullest pleasure—pleasure, that is, that does not depend on ignorance—is perhaps the profoundest enactment of our connection with the world.

— WENDELL BERRY, "THE PLEASURES OF EATING" (1990)

The train takes two hours, London to Swindon, then the taxi driver "thinks" he knows Bishopstone. "If we get lost, we'll just call, mate," and off we go. It's not as if I have come to the middle of nowhere, just to the rural English countryside, the kind of place Americans imagine when we imagine England—fields with hedgerows and loads of sheep. But, as I learned in London at CPRE, a lot of that England has been lost, paved over with housing and highways. Even the countryside that's left isn't the countryside as it was; many of the hedgerows were torn out in the 1960s and 1970s to consolidate farms and make fields larger to accommodate an increasingly mechanized style of agriculture. It's a style that depends on chemicals to fertilize crops and to kill unwanted creatures, a style that America knows well, one that relies less on farmers and farms than on "producers" and computers and oversized, air-conditioned, cushy-seated, stereo-equipped, GPS-guided combines. It's a style of agriculture that separates us—whether we actually farm or not, and almost no one does anymore—from the ecosystem on which all life depends: soil.

But I am heading now to a farm that follows a different style—Eastbrook Farm—and I see the sign before the driver does: "The Royal Oak," the pub associated with the farm, and her name, "Helen Browning," owner of said farm and pub, the woman I have come to see. "Well-kept ales, great company, fantastic organic food," says the sign, "using local organic, foraged, wild and seasonal ingredients throughout." And then, as though that isn't already enough to make anyone happy, it concludes, "Enchanting beer garden, roaring log fires." I believe I have come to the right place. Relieved to hear me say so, the driver gives thanks for the too-generous American tip and hurries away, leaving me to wander toward the one-story headquarters of Helen Browning's Eastbrook Farm.

"I have no idea why you're here." That's the first thing she says after bestowing a smile. Our connection has come through a friend of a friend telling me about Helen's work, how she not only runs the business of Helen Browning's Organic but is also CEO of Britain's Soil Association. She readily agreed to have me visit the farm, but she has a point. I mean, what more is there to say about good food and eating well? Michael Pollan has done three books on the subject so far, and he's just the tip of that spear. But what hasn't been covered quite so much is the very basis for all that good food and eating well. In fact, while I can't wait to try the Royal Oak's "fantastic organic food," I am really here to talk with her about dirt.

Dirt—or, more precisely, soil—is quite literally the foundation for human life on Earth. Yet around the world, we are degrading and depleting soil at a reckless rate. One recent study estimates that if we continue our current pace of soil abuse, we have only sixty years of harvests left. Our lack of attention to soil is hard to overstate—there is no human life without it, we are in the process of wasting it, and almost no one aside from specialists seems too concerned. It turns out that soil—the living entity beneath our feet—is the most amazing world that we know almost nothing about.

After nodding her understanding that I have indeed come to her for a reason, Helen suggests walking the farm before supper, while we still have light. It sounds like the perfect first step. And if there will be a roaring log fire when we return, all the better.

With short brown hair and sharp facial features that remind me of the pop star Annie Lennox, Helen pulls on tall Wellington boots, leaves the blue Subaru wagon unlocked after our short drive to the trailhead, and calls her greyhound named, apparently, "Dog." As in, "Here, Dog! Dog, Dog, Dog." ("Yes, we call the dog 'Dog,'" Helen tells me later, "though his given name is Digby.") And we are off, marching up the hillside at a pace that quickly has my shins and calves screaming. At the top of the hill, we head out into a field of spongy ground with gaps between deep green tufts of grass and clovers, nothing like the carpet of turfgrass that makes up most American lawns. This is grass that supports a menagerie of sheep, cows, cattle, and pigs, and plantings of wheat, oats, and peas. All this life supported by what some call "the skin of the earth"—soil, or, as Helen describes it, "six inches of black on top of chalk."

The terrain slopes and slants like ocean swells in shades of brown and yellow—and green in all directions. Hedgerows divide fields, and trees bunch close here and there. In early spring, the breeze keeps us company; the only other sounds are our footsteps and the faraway bleating of sheep. In the distance stands a small mansion owned by pop star Pete Townshend. Otherwise, the only lodgings visible are at the far end of the fields, the small "arcs" that the farm's pigs call home.

We head toward them, Digby the Dog leading the way. When we pass the herd of black cattle, they come trotting over to greet us, the happiest cattle I have ever seen. The sheep, some 350 ewes, are next. Helen's daughter Sophie and her fiancé own and tend the sheep. The ewes are less gregarious, backing away warily as we pass. But I probably would too, if I were one, as they are "about to pop" says Helen, and so have had their backsides shaved to make the birthing easier. Easier on the humans helping them along, I'm guessing. "Indeed," she says with a laugh.

I am laughing too, walking this farmland with its lush grass and soft ground and obviously happy animals. I muddy my shoes and snag my sweatshirt hood on barbed wire crossing through the fence, and think *I love this!* It reminds me of skinning my knee, that happy feeling of being a little kid again, wearing his scrapes and cuts and tears as little badges from a day well spent outside. Yet I also sense that I am just a visitor and that the ground here feels foreign, as it would to most of us now.

How many of us ever set foot on a farm? The grounds from which nearly all our food comes, do we know them well? I don't. Farmland: what does that even mean? And anyway, this is not what most American farms are now—somewhere distant, forgotten, awash in corn. Those are not places where you're invited to walk the land. But here—send Helen an e-mail and she will happily hike the grounds with you as she does with me.

There's knowledge that comes from walking the natural ground, and she has more of it than most of us. "When you walk the ground, which I do a lot, you get very used to feeling it," she says. "There's the old saying 'There's no such good fertilizer as the farmer's boot.' I don't think you can farm from a car or from a tractor; you have to get out and walk it. Because it's when you walk it you can feel what's going on." She nods back toward the hilltop we first walked. "To me that feels very different to walking a field where the organic matter is gone or the structure is gone. So that sense of feel through our feet, it's not something I think about, because it's instinctive. I would never talk about it, but I'm very aware that that's what I'm doing."

She looks toward the horizon. "And the other thing I've always done," she says, "is I love to lie on the ground. It's a very different sensation to lie on earth, a different level of connection . . . you feel like you're part of it at that point."

Helping visitors understand that we are "part of it" feels like the whole point of this farm. Aldo Leopold described "it"—the creation, the life of the planet—as "the stream of energy which flows out of the soil into plants, thence into animals, thence back into the soil in a never-ending circuit of life." From dust we came and to dust we shall return, as they say. Or, perhaps more appropriately, from microbes we came and to microbes we shall return.

I began this book with the idea that the oldest spiritual traditions and the newest science both tell us the ground is alive. But I had much to learn. In the past few years there has been an enormous expansion in our knowledge about the microorganisms below ground that are the foundation for life above ground. In fact, simply in terms of numbers, the wilderness below ground blows the world above away. For example, a tea-spoon of healthy soil might hold a hundred million to a bil-lion individual bacteria. That sounds like a lot to me, but also like a number that doesn't mean that much. Perhaps more startling is the fact that every living being depends on the bacteria and other microbial life in the ground. As one wild-life biologist told me, "If everything else is eliminated, and only microbes are left, there will still be life on this planet. But not the other way around. You take away microbes, we're all done."

The ground—at least where it's not covered in con-crete or asphalt or saturated in chemicals—is very much alive, the soil an ecosystem full of creatures about which we have almost zero knowledge. Plant roots, bacteria, fungi, protozoa, algae, mites, nematodes, worms, ants, insects, and grubs all make their homes in the soil. You want diversity? A single square meter of forest soil may hold more than a thou-sand species of invertebrates. The UN's Food and Agricul-ture Organization reports that "a typical, healthy soil might contain several species of vertebrate animals, several species of earthworms, 20–30 species of mites, 50–100 species of insects, tens of species of nematodes, hundreds of species of fungi and perhaps thousands of species of bacteria." They use that word "perhaps" there at the end because it's esti-mated that "we have identified fewer than 1/10th of 1 percent of the microbes in the soil."

As I walk with Helen on the soft six inches of living soil separating us from the dead chalk below, I think of my visit with Noah Fierer, a microbial ecologist at the University of Colorado in Boulder, and how he compared the state of his field to early-nineteenth-century botany, saying, "Basically we're just trying to figure what's out there." It reminds me of Thomas Jefferson sending Lewis and Clark off with the hope they might find mammoths still wandering the vast unknowns of the American West, and with instructions to bring back one of every plant they found. "There are so many knowledge gaps," Fierer had told me. "The questions we have are like, What's out there? What influences them over time? It's really just basic natural history."

Others share his sentiments. In late 2015, in papers published simultaneously in the prestigious journals *Science* and *Nature*, dozens of scientists called for a major initiative "to better understand the microbial communities critical to both human health and every ecosystem." Echoing da Vinci, one coauthor of the *Science* paper wrote, "It's like we're looking up in the sky with a refractive telescope for the first time and saying, 'Wow, it's amazing what's up there. What is all this doing? How does it work?'" Fierer explained trying to understand the microbial world to me this way: Imagine you were somehow able to grind up a forest into a tablespoon. Now try extracting from that tablespoon of grounds a picture of the forest based only on DNA sequencing. "It would be like trying to describe a forest to someone who had never experienced a forest," he said, "using only a list of the species that were found there."

Fierer and his colleagues realize that in order to help people understand how important soil life is for humanity and the rest of life on Earth, they need to better explain the amazing biodiversity in the ground beneath us. The fact that we live in such a visually oriented society, and that we usually can't see with our own eyes the life in the soil—besides occasional charismatic megafauna like earthworms—makes that a challenge.

In the summer of 2012, a team of scientists led by Fierer and Colorado State's Diana Wall, a world-renowned authority on nematodes (a microscopic soil animal of which twenty-five thousand species have been classified so far), came up with a novel way of grabbing the public's attention. First, they thought, they would need to find a good location, and then they would need to explore. "We were like, all right—Central Park's ideal," Fierer explained, "because everyone knows Central Park. You can look at an aerial picture of it, and you know when you're in the park and you know when you're not." On a hot July day, with a team of volunteers, in the middle of the nation's largest city, they set out to sample the biodiversity of the soil.

The results? Even they were astonished by the incredible diversity of soil organisms they encountered—as diverse a group as they might have found in a tropical rainforest or boreal forest. And almost every organism they found was previously unknown. "I think that project highlighted a point that if you're looking for novel organisms, you don't have to go to a deep-sea trench or the middle of the Congo," Fierer told me. "You can just go to Central Park and look at the soil."

We are only just now beginning to understand the vast life in the soil, what it does, and how our activities on the surface may affect it. That scientists like Fierer and Wall are today raising the alarm about the impact those activities have on the life in the ground makes the founding of Britain's Soil Association more than seventy years ago all the more amazing. Even then, in 1946, the Soil Association's charter members were concerned with "the loss of soil through erosion and depletion, the decreased nutritional quality of intensively produced food, the exploitation of animals in intensive units, and the impact of large intensive farming systems on the countryside and wildlife." Today, those at the Soil Association believe, "in the face of climate change and a growing world population, business as usual in our food and farming system is not an option." They identify themselves as "the UK's leading membership charity campaigning for healthy, humane and sustainable food, farming and land use."

As you might imagine, most of this language comes straight from their website, along with plenty of information about their campaigns on behalf of bees, seeds, antibiotics, and more. To actually be with their CEO, walking her farm, I can see how these values are being put into action.

The 1,350-acre Eastbrook Farm has been in Helen Browning's family since the early 1950s. Soon after assuming control in 1986, Helen began transitioning to organic production, and in 1989 established Eastbrook Farms Organic Meat. By 1994, the entire farm had gained organic status, and Helen began to increasingly focus on the connections between healthy treatment of the land and animals and the resulting health and happiness of the humans who ate the food.

As we continue our walk, I can't help but grin as we hear squeals from piglets and see them racing around like packs of puppies. The pigs here are British Saddlebacks, with black heads and hindquarters and a white stripe over their shoulders, their long ears folded forward almost like blinders. At any one time there are around three thousand pigs here, aged one day to ten years. We stand a long while as the piglets run to and fro, rummage the ground, tumble over one another, and pester the sows for milk. The sows are so much larger than the piglets, they could easily kill one if they laid on it, and so I'm impressed by how the sows lie down on their sides ever so slowly to avoid crushing them. For the growing number of city and suburban dwellers — including me — being in the presence of farm animals is nearly as beguiling as being near wild animals. I can almost imagine a future "English Farm National Park" where people will lean from their electric cars to photograph cows and sheep and ducks and chickens. So exotic! I find myself chuckling as we watch. They don't seem at all like trapped beings to be rescued but like happy farm animals doing what they do.

Of course, these pigs will be killed for food, nearly all of them. The piglets get "two months with mum, then four months with their mates" in a different field, away from the sows, Helen explains. So only six months of life before slaughter. But compare this to our treatment of pigs in the United States, where we take these intelligent, sentient creatures and raise them pumped full of antibiotics, packed together between iron bars and on concrete floors, rarely allowed outside to root in the natural ground. This so that consumers can have cheaper pork and corporations higher profits. I can't think of any better symbol of our separation from the grounds that give us our food than our treatment of pigs.

But it does not have to be that way. "As long as they've got soil to root in," Helen says, "they are just happy." And if people are going to eat meat, she believes, "then they have every responsibility to make sure it comes from somewhere happy." Raising and killing these pigs is something she has wrestled with "a lot," she says, "just the sense that I have the blood of hundreds of thousands of pigs on my hands." But, she continues, as long as people are eating pork, she feels it's her responsibility to give pigs "as good a life and as good a death as is possible to have in a farming system." Still, killing a lot of animals is part of the job.

Then again, she points out, "We also kill a lot of animals when [we] build a new road, or when we turn our central heating on. We may not have the knife in our hand, but we are changing the environment such that a whole bunch of other stuff can't live. And so in a way it's a bit more honest to be killing things saying I'm going to kill you and I'm going to eat you, than it is to do it with your eyes shut, as you go about your self-indulgent life."

What does all this have to do with our relationship to the ground? Quite a lot, actually.

First of all, pigs interact with the ground in a special way. "Yes," Helen says, "they destroy it. I mean, pigs like to root, dig with their noses, search for worms and beetles and other buried treasure. They have a huge number of nerve endings in their snouts, and," she says with amazement, "they just keep digging all the time."

Second, in churning and digging, in doing what pigs like to do, they contribute to the health of the ground. "The fertility they put down is tremendous," Helen explains, "and we always grow really good crops after pigs."

Third, it works the other way, too—the ground is good for the pigs, she says. The microbes in the soil help them maintain their immunity. The usual argument for housing pigs in concentrated animal feeding operations—CAFOs—confined in stalls and kept inside, is "efficiency," which means that for the amount of grain you put in the pig, you will get more meat. But should "more meat" be our main consideration? "Pigs are like us," she explains. "If you're running around, you might look healthier, you might be happier, but you will burn off more calories. So our pigs won't be as feed-efficient, and that's why organic pork costs a bit more."

That's part of it, but organic pork really costs more because it is raised in a way that honors the cycle that Aldo Leopold identified—from soil to pigs to humans and back through our treatment of animals and the planet.

Before my visit, I hadn't realized that the Saddlebacks were so much the heart of this farm. But everything I would soon learn about the way we are treating soil would make sense in light of my new knowledge about pigs. The way most of our food is produced is farming based on the artificial separation of soil from animal and of food from source, whereas Eastbrook is farming based on cultivating natural connections. It is farming based on walking the ground and knowing it well—its feel, its creatures, its life. The badgers and hawks, the pigs and sheep, the daughter and the town pub. "Humanity's future depends on the way we treat our soil," Helen says as we head back to the car. "And people are a long, long, long way from recognizing that."

For all the movement toward an appreciation of "local" and "organic" food at our restaurants, groceries, co-ops, and farmers' markets, we still have a long way to go in understanding the role—and therefore the health—of the local ground in which this food is grown. We can come at it from what the French call *terroir*, defined as earth or soil, with a specific meaning when applied to food or wine. Terroir is short for *goût de terroir*, meaning "the taste of place," and has to do with the way a particular environment—especially the soil—imparts a particular flavor to whatever animal, vegetable, or grape is being grown. Terroir is a key component of what Helen Browning is selling. The fantastic taste of her bacon and sausage comes in large part from the distinctive ground on which they are raised.

Perhaps for our grandparents or great-grandparents growing up on a farm, this would seem self-evident. But for those of us alive in the first part of the twenty-first century in America, the idea of terroir is novel and speaks to a larger truth. As Rowan Jacobsen writes, "We are some of the first people in history not to have built-in connections to the land we inhabit." In our increasingly paved world, terroir takes on a special meaning, offering us a way back into the ground. "Paying attention to terroir is one of the best and most enjoyable ways to reestablish the relationship" to the land, says Jacobsen. "It can teach us much about who we are...and how we go about living on this earth."

But first, we have to know more about the soil.

Soil

A thing is right when it tends to preserve the integrity, beauty, and stability of the biotic community. It is wrong when it tends otherwise.

— ALDO LEOPOLD, "THE LAND ETHIC" (1949)

Weaving through southern Virginia on a rainy weekday morning, I am on my way to visit Kristin McElligott. I first met her months ago at the annual conference of the Soil Society of America, where she was dutifully standing by her poster explaining her research into loblolly pine management. I liked her immediately—she was superbright, talkative, and funny (she described the forest's designed purpose as "to produce a crapload of loblolly pine"). The drive south from Harrisonburg takes me two hours, first down I-81 and then along winding country roads. Lots of yard signs like this: "America: Pray and repent. Return to Jesus." Lots of American flags. And lots of yard signs blending Jesus and American flags. A soil scientist at Virginia Tech, Kristin is familiar with this deep-red part of the country. Her previous degrees and work experiences have seen her living in Arkansas, Idaho, and Alaska, and she was raised in small-town Wisconsin, the first in her family to graduate from college. Given a choice, she admits, she would spend time tromping through wilderness areas in national forests or parks, but working in tree plantations has taught her much about balancing resource use and protecting the wilderness in which she feels most at home. Picture her with long brown hair in a ponytail, legs in jeans planted in hiking boots, striding off to find her research plots, her happy dog Nyssa running alongside. While the wet winding drive hasn't been much fun, I am immediately glad I have come. This woman is *passionate* about soil.

"I like to think of the whole belowground ecosystem as a distorted mirror of what you see above ground, just on a different scale." She has a spade in her hands—called a sharp-shooter down south—and we are walking through stands of loblolly pine toward a ditch she wants to show me. "Most people look at the ground and they see the surface. Soil scientists like to dig as deep as we can to figure out what's happening, how nutrient contents are changing, how microbial communities are changing, where the fungi are, where the roots go, where the water moves. All of that happens below ground."

This is probably the first thing to learn about soil: it can be incredibly dynamic. Unless it's been scraped away, covered, trampled, or otherwise killed, the soil under our feet can vary in consistency from one plot to another close by, as Noah Fierer and Diana Wall found in Central Park. This matters, because it tells us that all ground is not equal in its ability to sustain life.

In fact, around the world the ground holds several thousand different types of soil, and some are far more fertile than others. Made of a mix of minerals, organic matter, air, and water, soils are often described simply as "clayey" or "sandy" or "loamy." But because soil formation is influenced by many factors, including the "parent material" (such as rock), temperature, rainfall, organisms, and time, different types of soil are probably at least as varied as the landscapes they support, making the classifying of all those types extremely difficult. Nonetheless, the US Department of Agriculture begins its soil taxonomy system by dividing soils into twelve main types, each ending in "sol," derived from the Latin word (*solum*) for soil or ground. Of these it is the Mollisols (*molli* meaning "soft") that are most valuable. These are the soils of the most productive farmlands in the world—the mid-latitude grass-lands of the United States east of the Rocky Mountains, the steppes of Russia, and the pampas of Argentina. And here is where the States won the soil lottery. While globally, Molli-sols represent only about 7 percent of soils, in the States they make up more than 21 percent. In fact, about 35 percent of US soil is either Mollisols or Alfisols, an only slightly less fertile type. In China, by contrast, only about 12 percent of the soil is usable for farming or grazing. That the United States has long been the world's breadbasket probably has less to do with American ingenuity than with the fact that we are living on much of the richest soil on Earth.

Kristin, Nyssa, and I have stopped where a backhoe has carved a six-by-six-foot ditch that gives researchers a glimpse of the soil's makeup here on the pine plantation. Each differ-ent type of soil has a unique character that soil scientists describe by examining its layers, or "horizons." And when they describe it, they often sound like this: "There are more ephemerals, so there's a lot of rapid root turnover, and those dying roots are broken down by microbes, and that creates a really thick A horizon. But these are pretty thin, and you get into the B and the C horizon very quickly. Then you have more clay accumulation. There is a saprolite layer, that's rock, so that's pretty much just schist bedrock parent material that has been broken down and heavily chemically weathered..."

Kristin pauses, perhaps noticing the glazed look that has no doubt descended across my face. "I don't know how familiar you are with soil taxonomy..." she says, then laughs.

Not very, I tell her.

"It's easier to show you," she says, and invites me closer to where the backhoe has sliced open the ground. If you've ever seen a roadcut, you've seen an exposed soil horizon. There's the top layer from which the grass or plants or trees emerge, and below that what look like the layers of a cake descending before gradually giving way to the underlying bedrock. This vertical cross-section view of the ground is something most of us see only as we drive to work or school or vacation.

"Let's start from the top," Kristin says. In this pine plantation forest, the first layer is the O layer, as in organic. This is the stuff anyone who has tromped through the woods knows well, such as rotting leaves, last year's plants, and pine needles. Beneath this organic layer you reach what's known as the mineral layers (for the presence of sand, silt, and clay) beginning with the A layer and descending through E, B, and C. Beneath this lies the R layer, better known as bedrock. I think again of the stunning fact that in many parts of the world, the distance between surface and bedrock is mere inches. It is the distance, it's not too much to say, between life and starvation.

For Kristin and other soil scientists, the most exciting layer is the A horizon, or more precisely the intersection of the O and A horizons, where the organic and mineral layers interact. "This is where all the action is," she explains. "This is the life of the soil, where most of the microbes are. And microbial happiness yields biodiversity." It's in this part of the soil where we see mycelium, made of fine white roots that look somewhat like a spider's web woven with extra-large-sized filaments. The rhizosphere (meaning "related to roots") zone around the roots is where the plant and the microbial world intersect, and it's the hot spot for microbial activity because of the high biological and chemical activity caused by the exudate and sugars the roots excrete. Because the microorganisms that are nourished by these excretions are found in the soil's upper layers, microbial activity decreases with depth.

We know the A horizon by another name: topsoil — the ground on which human life depends. It is also, as Kristin says, "the most delicate, the most susceptible to degradation and erosion. The part that's right beneath your feet has the bulk of the life and the nutrients. It's what supports plant growth, microbial life, influences hydrology. Agriculture and development often degrade this part first."

Knowing how vital it is, it would make sense for us to treat topsoil with the greatest of care, even reverence. "Yes, but we haven't been convinced why it matters," she says. "I think if you don't see it, feel it, touch it, smell it, then there is no way to have the kind of connection with soil that is needed to understand its importance. Clearly we need to use it — it's an essential resource — but we need to use it with sound ethics. You have to be convinced that it's important and worth protecting."

She pauses, choosing her words. "I prefer to do that without fearmongering. But at the same time, it's a critical issue, and without healthy soil, societies collapse."

It's hard to believe that American society could possibly collapse because of a lack of soil. And it's true that we in the

States are blessed to live in a country so rich in this life-giving source. But in a small world growing smaller all the time, what happens to the soil in other parts of the world—often much more at risk than our soils—will eventually affect us and our economy, and the stability of the world around us.

For example, soil scientists fear that we are wasting and damaging our topsoil—again, the layer in which most of our food grows—at an entirely unsustainable rate. How unsustainable? On average the world has only sixty harvests remaining, reported one recent study. "On average" because although in the United Kingdom that number is one hundred harvests and in the United States the number is even higher, for other parts of the world—think Africa, India, China, and parts of South America, where the human population is largest and growing ever larger—the number of remaining harvests is lower, meaning that in fewer than sixty years the topsoil will no longer support the growing and harvesting of food.

Two incompatible facts: at the very moment when we know that by 2050 we will need significantly more food, we are paving over some of our most fertile soil. Human settlements have traditionally taken root in fertile areas, and as these increasingly urban areas grow in human numbers, we are developing this ground and thus losing the best soils for growing food. In the United States, the amount of ground being lost to development is stunning—more than a million acres a year. As one result, whereas in 1980 the nation had an average of nearly two acres of cropland for each citizen, thirty years later and with ninety million people added, that number had fallen to 1.2 acres per American. "How an Exploding U.S. Population Is Devouring the Land that Feeds and Nourishes Us," reports the subtitle of a study on sprawl. And once this ground is paved, there's no going back. As one expert noted, "Asphalt is the land's last crop."

While soil sealing and sprawl are urban-focused impacts that many of us can see at our feet, other serious threats to soil take place far from sight. These are primarily threats created by agriculture, and especially industrial agriculture as practiced by Western countries and exported to developing lands. The main culprits? Intensive tilling and the overuse of synthetic fertilizers and pesticides. The resulting degradation of soils includes compaction, acidification, and the decline of organic matter. Around the world, experts say, about 40 percent of soil used for agriculture is already considered either degraded or seriously degraded, meaning that in this 40 percent at least 70 percent of the topsoil is gone. In total, in the past 150 years, half the topsoil on the planet has been lost.

This means a lot less food for an already hungry—and ever-growing—human population. "Under a business as usual scenario," says John Crawford, an Australian sustainable agriculturist, "degraded soil will mean that we will produce thirty percent less food over the next twenty to fifty years. This is against a background of projected demand requiring us to grow fifty percent more food, as the population grows and wealthier people in countries like China and India eat more meat, which take more land to produce."

The potential for human suffering and environmental catastrophe is enormous. Consider the East African country of Tanzania, home to a human population of some fifty million. Tanzania is also home to an elephant population already decimated from years of poaching. In just six years, from 2009 to 2015, the country saw more than half its hundred thousand elephants killed. What happens if, as projected,

Tanzania's human population doubles to more than one hundred million in the next twenty years, while at the same time the soil's ability to produce crops diminishes? What happens to wildlife when millions of people don't have enough to eat? And then what happens when the wildlife is gone? Similar scenarios for disaster exist all over Africa, and on other continents as well.

Even in considerably more stable situations such as in North America and Europe, we are not immune from the consequences produced by the continuing loss of soil and depleting of soil's quality. For example, degraded soil means soil that contains fewer nutrients and grows food that is less nutritious. That's why, as Crawford explains, modern wheat varieties have half the micronutrients of older strains, and the same is true for fruits and vegetables, many of which have lost a significant percentage—sometimes more than half—of their nutritional value just since 1950. "If it's not in the soil," he says, "it's not in our food."

All this might not matter so much if we could just find more soil, or just make soil ourselves. But for all practical purposes soil is a nonrenewable resource. The recipe for soil is incredibly complex, requiring an intricate mix of the right chemistry, biology, and physics. And it simply takes a long time to form. The rule of thumb? Between five hundred and several thousand years for an inch of topsoil.

Which brings us back to sustainability. What we need to be talking about, one soil expert told me, is, "Can we continue agriculture the way we've been doing it the past fifty years for the next two hundred years? The answer is almost certainly no."

In fact, while there are many examples of how our way of

life is unsustainable, our abuse of soil may rank as the worst. The British writer George Monbiot recently described our soil crisis this way:

> Imagine a wonderful world, a planet on which there was no threat of climate breakdown, no loss of freshwater, no antibiotic resistance, no obesity crisis, no terrorism, no war. Surely, then, we would be out of major danger? Sorry. Even if everything else were miraculously fixed, we're finished if we don't address an issue considered so marginal and irrelevant that you can go for months without seeing it in a newspaper.

The number of harvests we have left, whether it's sixty or ninety or thirty, isn't the point. The point is that if we do not change the way we farm and build, we will run out of soil. "Almost all other issues are superficial by comparison," writes Monbiot. "What appear to be great crises are slight and evanescent when held up against the steady trickling away of our subsistence."

There is some good news. We actually know a lot about how to build and farm in ways that are less destructive to soils. "No-till" farming has been shown to successfully grow crops without the destructive force of tilling; organic farming methods can start to address the overuse of chemical fertilizers that have turned much of the world's soils from fertile ecosystems into sterile holders of plants; and the planting of cover crops can return carbon and nutrients to the soil. "The most important lesson is to keep the soil covered," Kristin McElligott told me. "When do you ever see bare soil in nature, naturally?"

So why do we continue to farm in a way that treats soil so destructively?

The standard reason given by the agriculture industry is that in order to feed a human population of seven billion and growing, there simply is no other way to farm. But while no one questions that feeding so many people is an enormous task that will only grow more challenging, this position seems to ignore reality. Explained one agroecologist, "People worry about what will happen if the oil runs out. But they don't seem to worry about what will happen if we run out of soil."

After talking with dozens of "soil people" and reading dozens more, I begin to think that the destruction of the soil upon which we depend has much to do with how we see the ground beneath our feet. And that is why I find myself heading to the state that represents the heart of American agriculture.

Ames

If we do not allow the earth to produce beauty and joy, it will in the end not produce food either.

— JOSEPH WOOD KRUTCH, "CONSERVATION IS
NOT ENOUGH" (1954)

I have always thought of Iowa as beautiful country. Perhaps it's because when I was a child my family often drove through the state to my grandparents' home in southern Illinois. Since those long-ago days I have made countless drives through Iowa. Time and time again I have passed Ames, the small city north of Des Moines where my mother was born, and time and time again I have thought of how my grandparents on both sides called northern Iowa home, how my father's parents lie buried in Corwith, just below the Minnesota state line. I have heard city friends complain about what a boring drive it is—the "endless" cornfields with "nothing" to see. But I have always felt a certain peace there, as though passing through some geography my soul still recognizes.

I realize I have been fooled. I have been seduced by images of small farms with barnyard animals, of small towns and caring communities, of dirt roads under stars. The truth here—as in so many places—is a far cry from what I have believed. The fields of corn I once found inviting now make me squirm. I know now about monocultures and pesticides, about nitrates and fossil-fuel fertilizers, about the dwindling number of actual families actually owning farms, and the growing number of enormous "farms" tilled by tenants on rented land, of landowners so removed from actual soil they could wear a three-piece suit all day and come home clean.

Maybe most of all I have learned how, with 98 percent of its original pre-European-settlement landscape gone, Iowa is by far the most transformed state in the Union. River bottoms drained, woods uprooted, and endless prairie nearly ended—in a state once 85 percent tallgrass prairie, only one-tenth of 1 percent of that prairie remains. In a very real sense, Iowa has been paved by corn—more than 82 percent of the state is devoted to cropland, and 50 percent of that is corn. In fact, so much of Iowa is planted with either soybeans or corn that one biologist I talked to described her state as "a giant open-air factory owned by monopolies."

I decide to visit Ames and spend some time with Fred Kirschenmann. Born in the 1930s in a farmhouse on the North Dakota prairie, Fred has been a farmer, an academic, a writer—sometimes all at once—a PhD in philosophy with years of experience having his feet on the soil. For five years he was executive director of the Leopold Center for Sustainable Agriculture, which since its creation in 1987 has been located at Iowa State University in Ames and given the mission by the state legislature to "research agricultural alternatives and share information about practices that also would increase the profitability of Iowa's farming operations and enhance the vitality of rural communities." Now retired from active farming, Fred divides his time between Ames and upstate New York, where he is president of the board at the Stone Barns Center, a farm-to-table operation led by renowned chef Dan Barber. Although Fred knows better than most people the enormous agricultural challenges we face, he comes to the conversation good-naturedly and with an easy laugh. More than once while discussing overwhelming and disheartening realities he will pause and say, "Again, I always try to look for a little good news," and then offer evidence that while the challenges are enormous, there is reason for hope.

By now many of the problems created by our industrial agricultural system are well known. Thanks to Michael Pollan and many others, we know that our world is made of corn, our food is often manufactured rather than grown, and high-fructose corn syrup is a four-letter word. We also know—I'm thinking of Jonathan Safran Foer's remarkable book *Eating Animals*—that the way we raise and slaughter our food animals verges on the obscene in more ways than one, with profits going to the few and costs paid by the rest of us, by the earth, and of course by the creatures themselves. Fred Kirschenmann is well versed in all of this, but his focus is especially on the fate of our soil. In a world where we may have only a century of harvests remaining, restoring the biological health of our soil is, says Fred, "our greatest challenge and most important task."

The challenge of changing our destructive treatment of soil begins with the fact that we have created an economic infrastructure that focuses on specialization and economies of scale. "That's the playbook," Fred tells me as we stroll the ISU campus, "that everybody operates by." The problem is that this playbook has only one play: produce a limited number of crops in huge quantities. As with other such single-minded plans for success, this may work well for a while, but eventually obstacles emerge. The obstacle when it comes to industrial farming? Farms are biological systems. In order for farms to thrive, the life on which they depend—especially the soil—must remain healthy. Aldo Leopold described the concept as "land health," which he defined as "the capacity of the land for self-renewal."

"So for farms to function effectively in the long term, you have to restore the biological health of the soil," Fred explains, "so that you have those regenerative capacities, instead of having to depend always on the inputs. At some point, we have to have a system that's more oriented to biological health rather than an input system. That's going to be the thing that's going to force change. The longer we wait, the more difficult that change is going to be."

When Fred mentions "inputs," he's referring to the poorly kept secret of industrial agriculture in the twenty-first century: it almost doesn't need the soil—it just needs some soil-like substance that will hold its factory-produced seeds, fertilized by factory-produced petroleum-based fertilizer, as they grow into factory-like rows of monoculture crops. As a result, Iowa's nutrient-rich soil has been pumped full of chemical fertilizers in order to meet ever-increasing demands for more and more product. Were those chemicals to run out or become prohibitively expensive—as Fred and others point out they eventually will—there's no way that this soil, as rich as it is, will be able to produce as much product (also known as "food") as it does now.

Unfortunately, the monoculture system we have now is good for agribusiness companies that benefit from having large quantities of corn and soybeans to process, export, and ship, with more bushels meaning more fees. Among other things, this makes changing the current system difficult. As one expert told me, "Everybody's got a vested interest in maximum corn and soybean. The seed companies, the chemical companies, it's all based on volume. So volume is good. And any kind of cutback is seen as a bad thing cutting into their margins. Everybody's into this. It's really hard to make this ship change course."

And yet, Fred explains, this ship has to change course. "We're not managing our land for health, we're managing it for maximum production.... Everything is focused on keeping the current system going. So the major subsidies go to the farmers who continue to grow more corn and soybeans, not to move in a more diverse integrated system. If farmers had to pay the full price of raising those commodities, it wouldn't work anymore."

Just as a cheeseburger is as cheap as it is only because we are not made to pay the true cost—to the earth, or to our bodies— corn and soybeans are as cheap as they are only because those who produce them, from farmers to multinational companies, are not required to pay their full cost. Those externalized costs are paid by consumers. We pay with our tax dollars as well as with our health, the health of the ecosystems on which we rely, and—perhaps most troubling—the health of generations who will follow us.

"It is banditry pure and simple," wrote the British botanist Sir Albert Howard in the first part of the twentieth century. "The using up of [soil] fertility is a transfer of past capital and of future possibilities to enrich a dishonest present." Howard was particularly troubled by what he called the NPK mentality (nitrogen, phosphorus, and kalium, another term for potassium), which dictated that you didn't have to worry about the life of the soil, you just inserted these three inputs and then everything would be fine. This reliance on chemical inputs at the expense of the soil's health, wrote

Howard, is "a particularly mean form of banditry because it involves the robbing of future generations which are not here to defend themselves."

The historical examples of societies that have failed through their mistreatment of natural resources are many—I'm thinking of Jared Diamond's work in *Collapse*, for example, which includes among its stories those of the Anasazi, the Mayans, and the Easter Islanders. But to find a dramatic example showing the consequences of soil mistreated, we need only look back to the United States in the 1930s. When drought came to a region that had once been lush prairie but now was intensely plowed and the soil no longer anchored by the roots of long grasses, the area's high winds grabbed the loose topsoil and flung it in massive black clouds across the country. The result? The worst human-created natural disaster in the country's history, the Dust Bowl, which devastated the nation's Great Plains and drove away more than 60 percent of the human population.

In response, in 1933 Congress created the Soil Erosion Service. Two years later, the name was changed to the Soil Conservation Service, the sole government agency whose mission was to protect this invaluable resource. In 1994, however, the name was changed again, this time to the Natural Resources Conservation Service. When I ask Dennis Chessman, the California state conservation agronomist for the NRCS, if he thinks the change reflected a reduced focus on soils, he says yes: "We've diluted our emphasis. We went from soil conservation to doing everything people could think of." Nonetheless, the NRCS is charged with working with private landowners to address environmental issues on their properties. "And," he tells me, "a lot of the things we're working with people to address can be addressed by improving the health and functioning of the soil."

Talking with Chessman, who in his role talks with farmers almost every day, I learn a lot about the ways we think about the current situation with soil, and the challenges someone like him has in changing ways of seeing the ground. "Sometimes we assume that simply by educating people on the complexity and value of soil, that we're somehow going to get them to act differently," he tells me. "But from the grower's perspective, the system is working just fine. Yields increasing over the past twenty years. Commodity prices are up. It's only the folks that have the long-term vision that really get this soil stuff that we're talking about. The rest of them just say, 'Yeah, but…'"

When I wrote my book on light pollution I often found myself thinking that if only we could jump back twenty years, we would be shocked at the change. But the steady growth of light pollution has been just slow enough that we don't notice the differences from year to year, let alone night to night.

The same is true for soil loss, says Chessman. "If we're talking about erosion rates of ten tons per acre per year, that is absolutely imperceptible. We're talking in the neighborhood of a sixteenth of an inch. A farmer will never see ten tons of soil on an acre leaving his field. But if he has that every year, eventually, regardless of how deep his soil is, he's going to lose his topsoil. It's the frog in the pot of boiling water analogy."

So far, at least, Chessman and others have not been able to get national governments to see soil as deserving of special protection. While in the United States we have a Clean Water Act and a Clean Air Act, there is no comparable Clean Soil Act. And we live in a society where what's under our feet is considered personal property. "This land is my land," says Chessman.

Exactly. And rather than seeing soil as something belonging to us all, where one person's (or company's) mistreatment of the resource impacts the community, we leave the fate of soil to the landowner. With air pollution we understand the concept of living downwind, and with water pollution the concept of living downstream, but we don't yet see how the effects of soil pollution spread far beyond property lines.

This view of soil as private property was exactly the argument used by European governments to derail a recent effort to get the European Union to adopt regulations designed to mitigate soil loss. Like the United States, the EU has laws designed to protect its water and air quality, but almost nothing to protect its soil. While it has recognized soil degradation as a "serious challenge," one that costs member states more than $40 billion a year, almost no new national or regional legislation to protect soil has passed in several years.

Why? Governments don't want to be told what to do with their soil, and neither do farmers. "Few sights are as gruesome as the glee with which the NFU [National Farmers' Union] celebrated the death last year of the European soil framework directive, the only measure with the potential to arrest our soil-erosion crisis," explains George Monbiot. "The NFU, supported by successive British governments, fought for eight years to destroy it, then crowed like a shedful of cockerels when it won. Looking back on this episode, we will see it as a parable of our times."

How we change this mentality is a question with which the Leopold Center is deeply engaged. At Marsden Farm, on the edge of Ames, ISU researchers have set up test plots to gauge the effectiveness of diversifying crop rotations beyond the dominant corn and "beans" (you hear that a lot around here—no one says "soybeans") rotation by adding small grains, red clover, and alfalfa into those rotations. Remarkably, their studies have found that corn and soybean yields in the more diverse rotations—rotations that would also begin to address the other costs, such as soil erosion and biodiversity loss—exceeded the yields in the conventional system. And yet, when I ask Fred why more farmers don't adopt these diverse rotations, he sighs. "You can't take anything but corn and beans to the elevators anymore. They don't handle anything but corn and beans." This means that farmers could grow a more diverse mix of crops but they'd have nowhere to take them, no way to get crops other than corn and beans to market. In addition, the crop insurance that so many farmers rely on favors those farmers who grow monocultures year after year, "so again that incentivizes them to do more corn and beans" rather than rotate those with other crops.

In other words, it's a pain in the ass to grow anything but corn and beans. The whole system is set up for those two crops, and that's what American society has been telling farmers to grow. As Fred says, "For at least the past half century, we have been telling farmers that all we want them to do is to produce as much food and fiber as possible, as cheaply as possible. We have told them they must specialize and streamline their operations, and at the same time we have invented the technologies to do so. Then we told them they had better 'get big or get out.'"

This last directive is one of two for which we remember Earl Butts, the US secretary of agriculture under President Nixon, who through his policies encouraged farmers to grow and grow and grow some more—if you want to survive, you'd better add acreage. The other thing we remember him for is the idea of farming "fencerow to fencerow," a concept that has been especially destructive for wildlife. Previously, animals, insects, and birds were able to survive around the edges of farm fields as long as at least some habitat remained. A farm planted fencerow to fencerow erases that life.

So we have been telling farmers to farm this way for several decades, and now we want them to change? In the United States nearly half our farmers are over sixty years old. In order to stay in business, most of the country's surviving farmers have invested millions of dollars in equipment and infrastructure to raise corn and beans. Convincing someone who has spent his or her life building such a business to change course? A tall task indeed, even when it's a way of doing business that from the outside appears entirely unsustainable.

"A friend of mine talked to one farmer and tried to help him understand this couldn't continue," Fred tells me, "and the farmer finally agreed that this was probably true. And so my friend said, What about you? And the farmer said, 'Well, I'm not going to be around.'" Fred laughs. "It's that 'mine and now' culture that we operate out of now."

During his eighteenth-century visit to this country, Alexis de Tocqueville noted the American tendency to "focus entirely on individual pursuits and leave the greater society to look after itself." This attitude is alive and well in twenty-first-century America, and one of the most troubling ways it manifests itself on the ground in Iowa is the gathering perfect storm of an aging population of farmers, land values driven skyward by investment speculation, and the erosion of the family farm. I was shocked to learn that almost half (47 percent) of Iowa's farmland is rented, because farmers can't afford to buy it. On Wall Street, investing in land has grown increasingly popular as a way to ensure a better, more reliable return than the stock market. It's a tension that could result, Fred tells me, in a state of affairs in which rather than family farms, we have "large plantations owned by investors, with 'serfs' operating the land because they can't afford to rent it or buy it." Almost by definition, those tenant farmers are not going to care for the soil as would a farmer who knows it as their own.

It can all seem pretty overwhelming—another situation in which enormous multinational corporations have control of our lives, and we are somewhat helpless to do anything about it.

But just now, for a moment, I forget all that, because Fred and I have driven out to the Marsden Farm and walked among the test plots. I'm hearing crickets and red-winged blackbirds. A monarch butterfly flutters past. And suddenly I find myself standing next to a field of oats.

Now, I don't mean to sound like I grew up in the suburbs, but I am thrilled to be standing next to a field of waving oats. It looks like wheat to me—strawlike stems topped by golden tassels—and comes up to my waist. I am tempted to drop my notebook and dash into its midst.

"Walk in," says Fred, polite but amused. "It's not going to hurt anything."

And so I do, oats swishing against my legs, a sound and sensation I don't know at all. I have been to battlefields and cemeteries and national parks, and I'm equally moved by stepping onto ground planted with oats. Generations of my ancestors, from my Iowa grandparents back through my family's Swedish and Norwegian and German history, are rolling their eyes. Has our descendant really become so separated from the farmed ground that gives him his food?

The answer, of course, is yes, for me and for most of us. How many of us could tell any more than I can the difference between wheat and oats? How many of us know what soy-bean plants look like, or even that soy comes from small leafy green plants growing close to the ground?

It makes me think about what it means to be a farmer these days. My bias, as a descendant of farmers, is that these men and women are filled with love and respect and a sense of responsibility for the natural world. Of course, this is exactly the image of the farmer that the gigantic multinationals use to sell their products, making us think that everything we eat comes from a happy little family farm. But we know that's not true anymore, that most of our food now comes from enor-mous farms where "farmers" have far less connection with the ground than they used to. As Fred says, "The problem with the kind of systems that we've created is that it becomes increasingly difficult to have any—or even want any—kind of affectionate relationships. It's all about manipulating and using the resources to get the maximum return."

Today the top 10 percent of the nation's largest farms control a whopping 70 percent of the nation's farmland. Since 1900, the average size of the American farm has grown by more than 200 percent, and since 1935 the number of farms in the country has fallen from more than six million to fewer than two million. While it's technically true, as the industry argues, that more than 90 percent of our farms are still "family" owned, these families may have almost no rela-tionship with the actual land. As example, Fred cites people he knows in North Dakota who have farms that have now grown to more than twenty-five thousand acres, and where some of the land is a hundred miles from the farmstead. "They're the first ones to admit that the only time that they go out to the land is in the spring when they plant and in the fall when they harvest. They have no connection."

Still, at the same time, something else is going on in our country. Even as the larger farms have grown larger, there has been a steady growth in the number of very small farms of fewer than one hundred acres. While these small farms are not yet in a position to replace the farms that supply our nation with food, their owners' more intimate relationship with the ground may point a possible way forward.

"I do think that continuing 'get big or get out' and 'fencerow to fencerow' is going to increasingly become dysfunctional as energy costs get higher, as the input costs get higher, the climate becomes more unstable," Fred says. "People are going to have to look more at how you put together resilient, regenerative systems. And that's going to move us toward smaller farms, because you can't have that kind of affectionate relationship when you've got forty thousand acres."

In fact, argued Wendell Berry, "it all turns on affection." In a 2012 lecture to the National Endowment for the Humanities, Berry borrowed the phrase from E. M. Forster's novel of rural decay, *Howards End*, to describe the key to farming's future. "Land and people have suffered together, as invariably they must," Berry explained. "But this has not been inevitable. We do not have to live as if we are alone."

Driving I-35 north to Minneapolis, I remember when I used to drive through Iowa and see beauty in the long rows of corn, imagining the birds and animals that must share this farm country, the calm clear evenings when work is done and you're walking hand in hand with your lover, your children trailing close behind laughing, your bird dog racing from one good scent to another. I wish I didn't see something else now in that corn.

I think of Aldo Leopold, born and raised in the northeast corner of Iowa, who wrote,

> My first doubt about man in the role of conqueror arose while I was still in college. I came home one Christmas to find that land promoters, with the help of the Corps of Engineers, had dyked and drained my boyhood hunting grounds on the Mississippi River bottoms. The job was so complete that I could not even trace the outlines of my beloved lakes and sloughs under their new blanket of cornstalks. I liked corn, but not that much. Perhaps no one but a hunter can understand how intense an affection a boy can feel for a piece of marsh. My home town thought the community enriched by this change, I thought it impoverished.

I have no doubt that even seventy years later, hunters have an opportunity to feel this affection because they spend time in marshes, in woods and fields—they have intimate contact with the natural ground. But we all have in us the capacity for this affection, this love for the creation of which we are part.

I heard that love of creation in a recent letter to the editor that the Minneapolis *Star Tribune* had titled "Once there was variety; now, a sterile sameness." The writer was a woman who has lived to see the changes wrought by industrial agriculture since she moved to southern Minnesota in 1960. "There was much more variety in the landscape…we heard and saw meadow larks and pheasants, and clouds of monarch butterflies were a part of every spring and summer. Now what do we have? Corn and soybeans from horizon to horizon; hedgerows with their diversity of plants and animal life gouged out; wetlands drained, and herbicides ensuring the…

increasing sterility of our natural environment." That phrase "increasing sterility" gives me shivers, reminding me of a hospital—and the way I always cannot wait to escape back into the land of the living.

Across the border in my home state I stop at the first rest area. I want out of the car, my feet on the ground. I park as far from other cars as possible and walk east toward the tilled field beyond. At first I see only one, but then another, then dozens still, fireflies floating close to ground in a summer dusk, in the rough edges between farm field and rest-stop lawn. They have found the wild strip of remaining earth, and I have found them.

We stand here on the planet's richest soil, in places several feet deep, centuries of gathered creation. It is arguably the most beautiful soil in the world—deep, dark, rich with nutrients and life. I stand soaking in the sight of small lit life, "shaky fire-sparks" John Muir's brother called them, bits of wildness in the otherwise sterile sameness carpeting this part of the country. Their pulsing light has to do with mating, the males calling to females, saying 'Come find me.' The females respond, and life continues. Seeing their small fires, I respond as well, reminded of the incredible way that, no matter how ravaged the place, there is always some wild life hanging on, continuing the dance it has known since time began.

I can't wait to be home. But at the same time I wonder what I will find. It's one thing to consider our separation from the agricultural grounds that give so many of us our food, and it's another thing to suddenly want to know whether "the increasing sterility of our natural environment" has reached our own back yard.

Grass

*A child said, What is the grass? fetching it to me with full
  hands;
How could I answer the child? I do not know what it is any
  more than he.*

     — WALT WHITMAN, LEAVES OF GRASS (1855)

G rass is ground we Americans think we know well. After
the floors of our buildings and houses, and the pavement of sidewalks and streets, grass is the third most common ground beneath our feet. By "grass" I mean turfgrass. I mostly mean lawns. I mean more than forty million acres, or sixty thousand square miles, of lawns—think the state of Georgia, that much lawn, cut into swatches, patches, and strips, and spread into every corner of every state—with five thousand acres added every day. We have more turfgrass than any other country in the world; none of the others even come close. We have three times as much turfgrass as we do corn, and we have a lot of corn. Altogether this turfgrass— residential lawns, commercial "estates," golf courses, and more—is now America's single largest irrigated crop.

The costs of this cultivation are enormous. Consider that most American lawns could never grow without our steady help—our water, fertilizer, pesticides, and time. Nationwide, on an average day, we use two hundred gallons per person of drinking-water-quality water to keep our lawns looking the way we think they should. In an average year, we spend $40 billion on lawn care—more dollars than our nation shares in direct foreign aid.

Even if we aren't homeowners, our society's many acres of lawn affect those of us living nearby. The US Centers for Disease Control and Prevention recently found lawn-care chemicals in 100 percent of nine thousand Americans tested, with the average person carrying thirteen of the twenty-three pesticides. And there's no question that by making our lawns inhospitable to insects, we send shock waves up and down the food chain—birds, bees, and butterflies are especially impacted. Said one writer in the *Washington Post*, "We disconnect ourselves from wildlife habitat loss by viewing it as a problem caused by industry and agriculture. But habitat loss is in our back yard."

Driving north from Iowa on I-35, I think of all the back yards, front yards, and side yards I mowed while growing up in a Minneapolis suburb. Especially in my early teenage years I earned money for music and comics and baseball cards by mowing neighboring lawns, sometimes several a day, coming home slathered in sweat and inch-dark clippings. I loved the smell of cut grass, the symmetry of straight lines, the sense of accomplishment. In graduate school I sometimes joked that when all else was failing in my life, at least I could successfully mow the small lawn I owned.

But now, as I head toward my hometown, I wonder about our relationship with turfgrass, not only about its monetary and environmental costs, but also about the ways it reflects and shapes our worldview. Twenty-five years ago, Michael Pollan argued that our lawns were "a symptom of, and a metaphor for, our skewed relationship with the land. They teach us that, with the help of petrochemicals and technology, we can bend nature to our will." Reading the latest studies and reports, I wonder whether anything has changed.

One thing seems certain: it's unlikely that our love affair with turfgrass will end anytime soon. For many Americans, a "great lawn" remains part of our dream. For others, the grass of a city park or golf course creates an essential oasis in an otherwise paved world. But "Turfgrass: yes or no?" need not be the question. We like grass, and we're going to have it. Perhaps the better question is, as with pavement, "How much will we have, and at what cost?"

To answer that, I'll need to know more about grass than just how to mow it. Where to start? By standing with bare feet, even if only briefly, on the nicest grass in the country.

At least, I'm certain no grass is nicer. As with almost 90 percent of the lawns surrounding the houses, apartment buildings, and businesses in Minneapolis, St. Paul, and their suburbs, the baseball field where Minnesota's Major League Baseball team plays is planted in Kentucky bluegrass. But that is about where the similarities end. This grass, this plush living carpet spread before me, is laid with laser-ensured evenness, mowed and trimmed to a shipshape, weed-free seven-eighths of an inch, and carefully fed and watered by a team of full-time groundskeepers who cultivate an island of turfgrass that is, like a museum piece, so perfect that most of us are never allowed to touch it.

That I am here this morning is thanks to Larry DeVito. Head groundskeeper here after previous jobs tending Fenway Park in Boston and Dodger Stadium in Los Angeles, DeVito arrived before the stadium was built in 2009 and was present to greet the turf upon its arrival after being harvested, rolled, and shipped in refrigerated trucks overnight from its sod farm origins in Colorado. Before the sod's arrival its bed had been carefully prepared, first by construction crews tearing up an old parking lot and clearing out a thirty-two-inch-deep space. After checking for any "irregularities" such as swampy ground or sewer pipes (one was found under the spot where the present-day shortstop stands), they placed eighteen inches of fill sand, four inches of pea gravel, and ten inches of root-zone mix made of sand and peat. It's this carefully constructed bed upon which the turfgrass was laid and upon which the multimillionaire athletes run, dive, and daydream.

Keeping this ground soft is a big part of DeVito's job and the reason he wears old running shoes to work every day. "You do it long enough," he tells me, "you get a sense for how the ground should feel." To me, the thick carpet of grass feels as though it's a stage. Compared to older stadiums, this one doesn't yet have a lot of history, DeVito admits. (The Twins' two World Series titles took place during their twenty-five-year indoor stay at the Metrodome.) But then he walks me over to the shortstop's usual position. "Well, here was Derek Jeter's last All-Star Game," he says of the New York Yankees great. "And Jim Thome's home run, when it hit off the top of the highest flagpole, that was crazy."

I said earlier that, aside from both being made of Kentucky bluegrass, the turf in our lawns and the turf here on Target Field have nothing in common. And it's true that few people have heating pipes buried six inches below their lawns to melt the snow if spring is slow to arrive. Or that even fewer get the kind of attention Larry DeVito and his staff give this turf—not even, it turns out, Larry DeVito ("I have a lot of weeds at home"). But as I feel the cool soft turf of the short, dense bluegrass on the soles of my feet, I realize an important similarity: for many of us, our lawns are where a lot of life's best moments take place. And not just for baseball, or just for Americans; think of all the grass fields around the world that have been home to good moments in human lives.

It's when a professionally manicured, chemically fed, proudly pampered island of turfgrass becomes the ideal for countless acres of lawns around our houses, businesses, and apartment buildings that things get complicated.

If homeowners in Minnesota go to Google with a question about their lawns, it's likely that their search engine will find Sam Bauer's name as the answer. The University of Minnesota's extension specialist knows more about grass than just about anyone in the state. And no matter the question, his biggest focus? "Everything I do is to try to reduce the amount of stuff we're putting on our lawns," he says.

One of the first things I learn from Sam Bauer is that people like Sam Bauer don't actually use the word "grass" much to describe their subject. "I work with turf," he says with a laugh. "That's what I do." And how is turf different than grass? "Turf means that it's mowed." The more precise "turf" also defines Bauer's focus as distinct from the wild grasses like those in the prairies that used to cover much of Minnesota and the central United States. For another thing, there is variety in the turfgrass used in American lawns. "There are two main categories," Bauer tells me. "We have the warm-season grasses and the cool-season grasses. The cool-season grasses, there are about five species that we use for lawns in the Midwest here. That's going to be your bluegrasses, your rye grass, the fescues, the bent grasses. And then there are subspecies." In fact, when Bauer later takes me to the university's grass research center, he'll show me checkerboard test plots of fifty different varieties of bent grass being tested for golf courses and fifty different varieties of bluegrass being tested for lawns.

Even in the space just outside his university building, Bauer is experimenting. In several areas he has replaced high-maintenance bluegrass with lower-maintenance species, turned off the water, stopped fertilizing, and barely mowed. "I'll show you one right now," he says, and starts off around the building. There, behind a roped-off section of hillside, I see fluorescent green turf nearly six inches in height, lush and thick. "It gets mowed one time a year," he says. Twenty feet away, across a sidewalk stairway, another section of turf looks terrible—sparse, patchy, the hillside soil eroding. "This is Kentucky bluegrass, planted in the shade. They put a lot of water on it. They're reseeding it every single year." The difference is stark. "This is the wrong grass species, that's all," Bauer says, nodding at the bluegrass. "It's just the Kentucky blue that we're all used to, and it's not the right spot for it."

Bauer and his colleagues are keenly focused on finding varieties of turfgrass that require less fertilizer, less mowing, and—especially—less water. The reason? The average American family uses more than three hundred gallons of water every day, with a third of that going to landscape (i.e., lawn) irrigation. Incredibly, as much as half of this water is wasted due to evaporation, wind, or runoff caused by inefficient irrigation systems or, says Bauer, just too much water.

"We use way too much water on our lawns," he says. "If you looked at the history the last five years, at least in the Twin Cities, you would have never had to turn your irrigation system on until June. People start them up in April and run them on an every-other-day program, thinking that their lawn needs that. It really doesn't." NASA researchers agree. In a recent study of American lawns conducted using satellite data, they found that while lawns in some locations (desert, for example) needed steady irrigation to stay green and healthy, lawns in rainier places needed almost no water at all.

Getting homeowners and businesses to understand this, though, is a constant challenge, says Bauer. He cites as example the local suburb of Woodbury, which uses two wells in the winter months to supply its residents with water. In the summer it uses nineteen. "They know that automatic irrigation systems are to blame," he says. "And the residents don't care. They like their green lawns. Even when I come in and say you can still have your same green lawns and cut your irrigation by thirty percent, they don't listen. They paid money for their irrigation system, and by God they're going to use it."

Nonetheless, when it comes to turf, Bauer exudes enthusiasm. "I love turf," he says. "I'll do a presentation in front of a hundred master gardeners and I'll ask, 'How many of you hate turf?' and ninety of them will raise their hands. Because they see irrigation systems running in the rain, they see fertilizer being spread on sidewalks. But to me, that's not a reason to hate turf. Our irresponsibility is not turf's fault."

Back in front of the building, Bauer nods toward two different lawn areas that to me look identical. Pointing to the area he identifies as a tall fescue, he says, "The irrigation system is shut off here, and I fertilize it once a year." Then he points to the area of Kentucky bluegrass just across the sidewalk. "The irrigation system there runs every other day, and that's fertilized four times a year." Bauer admits that at times, whether from heat or lack of rain, the fescue doesn't look "as nice" as the bluegrass, which when you're looking for alternatives for people's lawns, presents a challenge.

"I probably spend four hours a day on homeowner questions," he tells me, "and the things that they're worried about in their lawn, I just can't believe. You've got to keep up with your neighbor—it's been ingrained for the last five decades. Everybody sees what your lawn looks like because it's easy to see the surface. But when you look at what everybody puts into these lawns, knowing the price tags that come with it, I think you would have a different idea of what you wanted your lawn to look like."

Seeing lawns in a new way—that's exactly what happened for Shay Lunseth, and becoming a new mother probably helped. When she met her future husband, Eric, he already ran a traditional lawn-care service, doing things "the way they had always been done," which meant using chemicals to keep Kentucky bluegrass alive. Shay started asking why there wasn't a more natural way to care for lawns. "When we started having children, we thought about the pesticides being put on our lawn," she tells me, and that was all it took. Shay went back to graduate school in horticulture, and Eric began experimenting with organic lawn-care techniques. Today, they run Organic Lawns by Lunseth.

With long red curls and a welcoming smile, Shay meets me on the driveway outside their house south of Minneapolis. The rear side window of her black Subaru shows a smiling baby sitting on a lush green lawn next to the slogan "Organic Lawn." The bumper sticker reads "Naturally, we can help." As we stand looking down at the front yard, I ask her what she sees.

"I see weeds," she says. "But we tolerate weeds." Unfortunately, the Lunseths have learned that some people simply cannot tolerate them. "We lost a townhome association contract this year because they were complaining about weeds," she says. "So I went out with the property manager, and they had 250,000 square feet, six acres of lawns, and we counted six dandelions. But they cited this when they canceled the contract." The only way to ensure that not even six dandelions show? "A chemical lawn," says. "A lot of people who want a nice lawn don't care how they achieve it."

"A lot" of people is an understatement. A solid majority of American homeowners, business owners, and others whose to-do lists include making the lawn look nice, pursue that goal by relying on petroleum-based synthetic fertilizers and pesticides. It's a goal set in motion by developers, who as a general practice scrape each new construction site of everything living—trees, plants, even the topsoil—in order to more easily grow their houses, buildings, and streets. The bright green lawns they then roll out are almost always non-native turfgrass. To keep this turfgrass alive, we spray, pour, and spread some ninety million pounds of fertilizer and more than seventy-five million pounds of pesticides per year. In doing so we create millions of sterile acres, ground where almost nothing can make a life, a green desert where, as author David George Haskell puts it, "sex and death have been erased." As a result, our mornings are less alive with birdsong, our backyards are rarely visited by butterflies, and we join in the decimation of pollinator species.

Information about the perils of pesticides and lawn fertilizers has been available for a decade plus. But it's hard to argue that we aren't still using way more than we need of both. For example, between 40 and 60 percent of the nitrogen we pour onto our lawns every year winds up in surface groundwater. In summer many a suburban lake sports green surface scum as a result—algal blooms that degrade water quality, steal oxygen from fish and other aquatic life, and threaten our drinking water—a process of eutrophication that impacts US water supplies to the tune of $2 billion every year.

Perhaps most troubling, our chemical lawns—especially the pesticides we use—have been tied to human health risks. Studies show that of thirty chemicals commonly used in lawn pesticides, nineteen are linked with cancer, thirteen with birth defects, and others with reproductive effects, liver or kidney damage, neurotoxicity, and disruption of our endocrine systems. The *Journal of the National Cancer Institute* found that "home and garden pesticide use can increase the risk of childhood leukemia by almost seven times." Argues Dr. Diane Lewis, author of *The Toxic Brew in Our Yards*, "We need to see a perfect lawn not as enviable, but a sign of harm."

Shay tells me that after years of spraying chemicals, Eric has recently been diagnosed with asthma. "It could just be the stress of owning a business," she acknowledges, "but this came out of nowhere. And he's an otherwise healthy person who's run five marathons in the last five years."

One thing troubling her now is Monsanto's development of Roundup Ready Kentucky bluegrass turf. Having already had great commercial success with Roundup—more than 90 percent of the corn, soybean, and cotton crops in the United States are genetically modified to be Roundup Ready, meaning that the product kills everything but the desired crop—Monsanto has created genetically engineered turf to be resistant to glyphosate, the "active ingredient" in the Roundup herbicide. This means that the modified plant, whether corn or cotton or the turfgrass of so many American lawns, is immune to being sprayed with glyphosate while almost every other living thing around it dies. Monsanto argues that "glyphosate safety is supported by one of the most extensive worldwide human health...and environmental databases ever compiled on a pesticide product." And that "when used according to label directions, does not present an unreasonable risk of adverse effects to humans, wildlife or the environment."

Perhaps. But this statement would no doubt be objected to by monarch butterflies, whose population has crashed since the use of Roundup began destroying the milkweed plants that are their primary food source. It would certainly be questioned by the World Health Organization, which in 2015 identified glyphosate as a probable carcinogen for humans. It would frustrate scientists who have found glyphosate in groundwater and drinking water. And it would confuse farmers who are seeing an explosion of Roundup-resistant weeds in their fields, a problem that "anybody working in agriculture" would say is "very serious," according to David Mortensen, a professor of applied plant ecology at Penn State University. In fact, the rise of these glyphosate-resistant weeds has led to the use of more herbicides, which in turn leads to more herbicide-resistant weeds, which in turn leads to even heavier use of herbicides, and so on. Already, glyphosate has become the most heavily used agricultural chemical in the world, with more than nine million tons sprayed onto fields worldwide since its introduction in 1974. Or, in other words, enough Roundup Ready herbicide to spray nearly half a pound on every cultivated acre of land in the world. Eventually, says Bill Freese of the Center for Food Safety, using ever-increasing amounts of glyphosate and other chemicals to control weeds "is going to be a dead end."

Indeed, the debate over whether glyphosate is safe or not

obscures the more important issue of our continued reliance on chemicals in order to have a "great lawn." Even with the documented costs to wildlife and environmental health, as well as concerns about the effects on human health, our overuse and misuse of lawn fertilizers and pesticides and herbicides continues to grow. Handing homeowners a spraygun full of a herbicide that has already ignited so much concern seems worth a significant pause. "Giving homeowners the opportunity to blanket-spray anything," says Shay, "is kind of a scary thought for me."

The fact is, by spraying different 'cides on our lawns, we are basically poisoning our lawns green. By using synthetic chemicals we kill off the upper level of microorganisms in the soil, and then need synthetic fertilizer to do the job the microorganisms might have accomplished. "So many people think that all the chemical companies out there are just putting down some kind of magic on their lawn that makes it green," Shay tells me. "They don't even know what they're talking about, they just want that 'weed feed' or whatever term they've heard on TV. They just want it to 'look good.'"

And, Shay explains, "We're not here to just make it look good." The Lunseths' organic approach to lawn care is to feed the soil with organic fertilizer, and then the soil will feed the plant. By doing so, they work to build soil health and to create an ecosystem that invites insects and soil biodiversity. If some weeds do appear, says Shay, she's not opposed to spot spraying—but, she says, "We can reduce the amount of chemicals by ninety percent by using Integrated Pest Management. Which is only applying the products when and where necessary. We don't need to spray this entire patch of lawn."

So you don't stick those little cards in people's lawns warning them to keep their children and pets off the grass for twenty-four hours? Shay laughs. "The chemical companies have the sign with the kids and pets circled and a red line running through. Our sign just has the kids and pets. You could sit in the grass as we're doing it, because there's no harm."

The value of turfgrass lawns isn't what's in question for the Lunseths. I'm reminded of the owner of a native-plant business in Virginia telling me, "There's nothing wrong with grass in limited amounts. We do need spaces to run around. We need spaces to lay the picnic out. But we've become *crazy* about having grass." For me, the best example of this craziness is the acres of turfgrass circling corporate buildings, offices, and business parks. "It seems completely unnecessary," Shay agrees. "It's a status thing. Maybe have a little bit in the front where your sign is, sure, but the rest could all be pollinator-friendly native plants."

And here is where she is most hopeful. The focus on bees and other pollinators in the past two years, she says, has increased dramatically. Her phone rings every day with potential customers wanting to "do something" with their lawns to help. In short, she says, "They're interested in making their lawn purposeful."

"A lot of people have not thought about it," says Lunseth. "The vast majority of people don't think about it. *We* think about it every day."

When I first began researching this chapter on grass, and especially when I began thinking about the idea of being purposeful with our lawns, I never thought the next place I would visit would be where I find myself this morning.

The first one to greet me is Josie the dog, a one-year-old shepherd mix from the local rescue place, Secondhand Hounds.

And next comes her owner, Jeff Johnson, superintendent of the Minikahda Country Club golf course.

I mean, here's the thing: golf courses are bad for the environment, right? Guzzling water, pumping out pesticides and fertilizer, ripping up native habitat to spread turf so green it almost seems unreal, engendering headlines like "America's 18,000 Golf Courses Are Devastating the Environment" and organizations with titles like "The Global Network for the Fight Against Fields of Golf." I admit, I fully expected to find these criticisms confirmed. But the story of golf is far more complicated.

With Johnson at the golf cart steering wheel and the exuberant Josie leading the way, we set off to explore the oldest and most prestigious golf course in my city. Named for the Sioux word meaning "by the side of the water," Minikahda was originally built at the end of the nineteenth century. While the course has been upgraded over the years, the topography retains the original hilly slopes the first members found when this was still countryside and not city. "They didn't have the heavy machinery then," Johnson tells me, "the earthmovers they would have needed to change it." Since the course's construction, Minneapolis has grown up around it, and so the first thing I feel here is the sense of going back in time. Johnson loves this feeling about the course—his office walls are covered with historic photographs—even preferring the more traditional titles for his job of "keeper of the greens" or "greenskeeper."

Keeping their courses green is a pressure under which superintendents all around the world labor, Johnson tells me. Some club members have the attitude "I don't care what you spend, I don't care what you apply, we want it perfect, just give it to us." Nonetheless, Johnson has slowly increased his efforts to make Minikahda a more environmentally friendly course. For example, rather than mow everything, as was done in decades past, more and more areas of the course have been allowed to go "natural." Now, Johnson regularly sees coyotes and foxes, has seen a tom turkey in "full gobble-gobble" attempting to seduce a hen, and says raptors love the burgeoning population of mice and voles the natural areas engender. The Minikahda course lies just a couple of blocks from the city's Lake Calhoun. As Josie races ahead, Johnson directs the quiet-rickety ride of the golf cart toward the 18th green, where we pause to look out over the lake to the east and toward the city's skyscrapers to the north.

I will admit I am predisposed to appreciate being on a golf course, and not because I've ever played a round of golf. In the suburb where I grew up, the local golf course has long been the biggest expanse of greenery and trees around. During the years when I would return home from graduate school out west with my bird dog, Luna, we would wait until after 11 p.m., walk the two blocks to the course, and slip through a tear in the chain-link fence. While Luna raced about with joy, I would follow the same path each time, finding even in the middle of a suburb a bit of wildness. We too saw coyotes and foxes and deer. In winter, the bare oak limbs would often host the small Buddha-like shapes of owls. Today, as our cities and suburbs have continued to expand, many golf courses are the greenest areas around. I still appreciate, coming back to my childhood neighborhood, the way the world is still wilder on the other side of the fence.

Nonetheless, the criticisms of golf courses for their environmental impact include an overreliance on heavy application of pesticides and fertilizers that often end up in the water and soil of surrounding communities, the destruction of native plants and wildlife habitat, and unnecessary water use. In fact, at more than 300,000 gallons, the average golf course uses the same amount of water each day as 2.8 million Americans combined. For its part, the golf industry points out that it adds more than $70 billion to the US economy each year, employs more than two million people, and has worked hard to alleviate its environmental impact. The industry knows it has a poor environmental reputation, and many courses have been making changes for the better.

At least, that's the way it seems to me. Without doubt there are courses, whether here in the States or abroad, that pay lip service to environmental stewardship and then ramp up their pesticide application at the slightest whiff of anything less than green. And clearly, to me at least, the destruction of native wildlife habitat so that wealthy patrons can dress in bright colors and zoom around in electric carts is wrong. But in my research visits I find consistent examples of superintendents balancing their customers' ideals of what a course should look like with efforts to reduce their environmental impact, and often with more sensitivity to their impact on the environment than the rest of us might have. Think of it this way: whether homeowners or business owners, anyone who has responsibility for a turfgrass lawn is a "grass farmer." I first heard this term from a golf course superintendent, and it immediately made sense. We're just like crop farmers, except we're not interested in yield, he told me. We just want quality grass. But a growing difference exists between golf course superintendents and most of the rest of us in how we achieve that goal.

Take water use, as an example. As we make our way back down the course, Johnson points out the different grasses in the roughs (bluegrass) and the fairways (bent grass and poa

The Bedlam burial ground was in use for nearly two hundred years spanning the 16th to 18th centuries, a period of time that included Shakespeare's plays, the Great Fire of London, and the Great Plague of 1665. A team of sixty archeologists worked in shifts six days a week for several months to remove and catalog more than 3,000 skeletons. (© Crossrail Ltd)

The Stone House was built in 1848 at the intersection of the historic Warrenton Turnpike and Manassas–Sudley Road and served as a makeshift hospital during the Battle of First Manassas. One visitor's account described the wood floor as crowded with wounded men "dreadfully mangled by cannon shot" and "their clotted wounds still undressed." Today, the busy intersection is where Virginia State Route 234 and US Route 29 meet. (Paul Bogard)

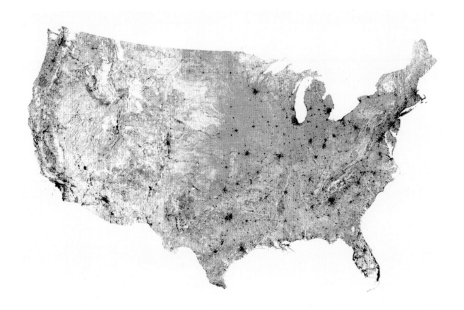

The creators of the All Streets map used data from the US Census Bureau to chart some 240 million individual road segments, but decided not to include Alaska and Hawaii because there aren't enough roads in either state to outline its shape. (Fathom Information Design)

The 82 percent casualty rate suffered by the First Minnesota Volunteer Infantry Regiment during their charge at Gettysburg is still the highest casualty rate suffered by any US military unit during a single day in US history. The inscription at the base of the statue reads, in part, "In self-sacrificing desperate valor this charge has no parallel in any war." (Douglas W. Orr)

Surviving members of the First Minnesota returned to the Gettysburg battleground in 1897 for the dedication of the monument honoring their charge. While there, they posed on these "witness stones" they remembered from the battle. The matching photograph is from 2010. (Perry Tholl)

Digby the Dog surveys his domain on the fields of Eastbrook Farm outside Bishopstone, England. (Paul Bogard)

A soil profile from Massachusetts showing A, E, and B horizons above bedrock. It is the narrow band of topsoil (A) on which all life depends. (Jim Turenne)

Samples from a loblolly pine plantation in southern Virginia demonstrate the incredible diversity of soil. Despite their differences in color, several of these samples were taken within a few centimeters of one another. (Kristin McElligot)

Sandhill cranes lifting into flight. Young cranes, called "colts," leave the nest within a day after birth and fly within two months, remaining alongside parents for nine to ten months. While on the ground, cranes will jump, run, dance, and, reports *National Geographic*, "otherwise cavort around." (Public Domain)

An aerial view of where forty acres of woods have been cleared to make room for a new fracking well pad in southeastern Ohio. (Ted Auch)

A fracking well pad at night. At the time this photograph was taken, in 2016, the six wells at this particular site had produced close to 300,000 barrels of oil. They had also produced more than 100,000 barrels of brine waste and consumed more than fifty million gallons of freshwater. (Ted Auch)

Symbolic train tracks at Treblinka, part of the memorial honoring the hundreds of thousands of people murdered here by the Nazis. At the height of the terror, trains delivered to the extermination camp more than 20,000 people a day. (Little Savage)

Permafrost on the Alaskan coast. Tens of thousands of years old, in some places more than 2,000 feet deep, this frozen ground lies beneath 85 percent of the state. (United States Geological Survey)

The author with Atty Gúndiwa Villafaña and Wilfrido Izquierdo at the Museo del Oro Tairona–Casa de la Aduana in Santa Marta, Colombia. They are Arhuaco, one of Colombia's indigenous groups and descendants of the ancient Tairona civilization that lived in this area beginning in 200 BCE if not before. (Paul Bogard)

annua) and the greens (bent grass). He says, "People say I must water a lot. But it's like if you sit on the couch all the time, you're just going to get fat and lazy. It's no different with turf—if you're feeding it and watering it and using pesticides all the time, you can condition that plant to need those things." Johnson tells me that computers track his water usage so he knows exactly how much each part of the course receives. "We're not just flinging water randomly," he says with a laugh. "And you're not going to drive by Minikahda and see irrigation going when it's raining. That's not the case when you drive by Target or any other big company. Or homeowners. They have no responsibility environmentally. We do, because we know the eyes are on us."

This is a sentiment I hear echoed in other discussions, and recent studies of golf courses as wildlife habitat seem to support this notion too. The lead author of a study on North Carolina salamanders admitted, "We went into the research study thinking...things were going to be really toxic and really bad...[but] what we found was quite the opposite—golf courses can actually provide a wonderful habitat for salamanders and other organisms." When it comes to how we treat the ground beneath us, an increasing number of voices argue, golf courses aren't the problem. The problem is more the rest of us who are not yet seeing the ecological impacts of the way we are farming our own grass, and the way we accept—not only as normal but as the ideal—a weed-free, chemically maintained green carpet devoid of insect and other life.

During our tour Johnson shares both a peaceful reverence for the course and a panicked concern at its condition. I hear it in the awed tone of his voice as he relates the story of finding a days-old fawn "nestled in the root flares of a big elm tree while mom's off foraging," and his almost embarrassed admission that he has laid down on one of his greens just to feel the soft bent grass beneath his back and watch the clouds passing overhead. But there's also "seeing everything that's wrong with this property and trying to fix it. You spend your day driving around thinking, *This needs to be done, that needs to be done.* You're striving for perfection all the time, and it's not attainable."

Most of all, I hear it in the way he tells me how well he knows the 155 acres under his care, the ground on which he spends his working life. "I know the property in and out. I know every green, every spot that will need a little extra water. I think I could probably go down every hole and know what the trees are," he says. "Oh, that ash tree on the right side of four? I know the one."

Jeff whistles and Josie immediately makes a beeline back toward us. "If you're here long enough, you learn."

In his book from 1994, *Becoming Native to This Place*, Wes Jackson writes that American universities "now offer only one serious major: upward mobility. Little attention is paid to educating the young to return home, or to go some other place, and dig in," a process that entails "becoming native to our places in a coherent community that is in turn embedded in the ecological realities of its surrounding landscape." About thirty seconds into meeting Julia Kay in the front yard of her south Minneapolis home, I realize for the zillionth time how far from this ideal I am too.

"I grew up in the eastern broadleaf forest biome, if you know the biomes of Minnesota?" She looks at me quizzically, and I admit that I don't. I quickly learn that there are four (the others are Laurentian mixed forest, prairie grasslands, and tallgrass aspen parkland), and that Julia grew up near to where they meet, "in a world where there was nature all around us."

Even without a working knowledge of my state's different biomes ("a large naturally occurring community of flora and fauna occupying a major habitat, e.g., forest or tundra"), I can count myself among the lucky kids who grew up in a similar world. My time in the suburbs mowing lawns and playing Little League were mixed with a generous portion of time spent at a northern Minnesota cabin next to a clear, spring-fed lake. My book about the night sky came in large part as a result of many nights standing in open-mouthed awe as the Milky Way curved overhead. This book comes from realizing how little I knew about the ground at my feet. And I want to know. What does it mean, as Jeff Johnson had said, to be here "long enough" to learn, or to take Jackson's phrase, "become native" to a place?

There certainly is nature all around us as we stand in Julia's front yard. For starters, her "yard" has no turfgrass. It is instead made mostly of native plants and insects. The tall and diverse mix of wildflowers stands out from her neighbors' more traditional Kentucky blue like a head of wild Einstein hair after a long line of army crew cuts. But around the city, more and more front yards are going native as citizens respond to the devastation caused by industrial agriculture and chemical lawn care to the pollinating insects that ensure the majority of flowers and fruits we humans enjoy. As Julia tells me, "The insects are the basis for the food chain. Without them, we don't survive."

It's a story worth repeating, for it's a story whose ending will depend largely upon us. A 2014 study published in *Science* reported that over the past thirty-five years, a period during which the human population doubled, the world's population of insects declined 45 percent. In the United Kingdom alone, the number of common beetles, butterflies, bees, and wasps fell an astounding 60 percent. If we need the threat of impacts to humanity in order to grab our attention, the losses of such insects should have dramatic effect. To start, about 75 percent of the world's food crops require insect pollination. Bee populations alone, which have fallen between 30 and 40 percent, are responsible for pollinating 75 percent of the nuts, fruits, and vegetables grown in the United States. Other "ecosystem services" provided by insects include pest control (in the States alone, an estimated $4.5 billion annually), nutrient cycling to ensure plant productivity, and human health (the loss of insects seems to increase the spread of human disease). And what we do know about how insects help sustain us is easily overshadowed by what we don't know. Although scientists have described about a million insects, estimates are that at least another four million more have yet to be described. In other words, we are losing insects and the benefits they provide without even knowing what we are losing. Says Jürgen Deckert of the Berlin Natural History Museum, "The decline in insect populations is gradual...and there's a risk we will only really take notice once it is too late." Thus far, national governments and world organizations have been slow to respond. Adds Deckert, "The key question is whether governments view biodiversity as an add-on or as something that is of existential importance for our future."

Meanwhile, people like Julia find themselves doing what they can with the ground they call their own. We walk around to her back porch, joined by the very personable "Dog" the cat. On the telephone wire to the house, a chickadee sings his heart out. A faux-antique sign that hangs nearby reads "This Is My Happy Place." Julia brings two glasses of lemonade from the kitchen. I ask about her neighbors, whether they appreciate her yard.

"I think there's a change in attitude in the younger generation. I think the younger generation is really scared."

"Of what?"

"The effects of climate, the effects of my generation and our parents not taking as good care of the land as we could have, the effect of the last thirty years of toxic food production. They're starting to really break the cultural norms of their parents."

There's hardly a stronger cultural norm in our society than having a "great lawn" of Kentucky bluegrass. The movement toward uniform lawns linking one house to another took off with the building boom in suburbs after World War II. In some Minneapolis suburbs, for example, developers actually passed out an instruction manual for how to make sure the grass in your yard achieved a traditional golf-course-like appearance. I know that when I was a child in the mid-1970s, there was no question about what the ground was supposed to look like: Kentucky bluegrass, from sea to shining sea. "The aesthetic is purely ornamental," says Julia. "It's all about appearances." Fast-forward to today, and the young families moving into those houses now need someone to provide an instruction manual for how to break that cultural norm.

One idea comes from a local organization, Prairie Restorations. It has begun a campaign they call "Sewing It Back Together"—linking the farm community with the urban community—by converting "25 percent of the mowed turf in North America to some kind of native plant community." It's an idea that Julia finds exciting. "We are a long way off from changing agriculture in the plains region, and the habitat loss out there is really sad. If we took 25 percent of everybody's yard and made a little natural corridor in our urban settings, then we have an oasis in the city that helps keep those populations of insects alive long enough for the agricultural world to catch up. Our cities can be habitat for all, not just for people."

Do we think of where we live as habitat? "The natural home or environment of an animal, plant, or other organism," reads one standard definition. Most other animals are uniquely adapted to their immediate surroundings. They live where they have always lived, and they don't move much (or move to the same place year after year). We humans are uniquely adaptable, able to live just about everyplace on Earth. But while the limited mobility of our fellow creatures ensures their intimate knowledge of their habitat, our fabulous mobility ensures our limited knowledge of where we are. Maybe now we have reached a point where knowing where we are is critical to where we—and our children— may be headed. As Wes Jackson writes, "Wilderness...will disappear unless little pieces of nonwilderness become intensely loved by lots of people."

When I ask Julia what she sees when she looks at a traditional chemical lawn, she says, "I see missed opportunities." Yes indeed. Rather than a chemical lawn, an organic lawn. Rather than simply turfgrass, native plants that invite native insects. In other words: all of these lawns that are all the same, with way more turfgrass than we need and with costs no one wants, are an opportunity to consider what it means to "become native," by which I mean to intensely love a little piece of ground we call our own.

*　　*　　*

Looking around, I'm suddenly struck by a very real question. "Where are we?" I ask. Julia ponders, then says, "Wait here, I have something to show you." In a few minutes she's back, unfurling a map before me. It's a map of Minnesota based on the original surveys of the state in the 1850s, she says—the entire geography divided into small squares—and shows me where the Twin Cities now stand. The red lines are original Indian trails; she says, "All of those trails are highways now." A few orange areas show where there was prairie, "and when you see the maps of the west part of the state, they are just solid orange." The white squares are woodland and brush, with tamarack the main tree. And this little yellow patch? "That was a marsh," she says, "and we're on top of that."

Talk about knowing the ground. As John Randel Jr. had done while surveying Manhattan four decades before them, mid-nineteenth-century surveyors of this land did their job "the old-fashioned way," by walking each one-mile section of each square. On the corner of each, they would indicate what they saw looking north, south, east, and west. What type of tree, whether they saw water, whether they saw particular plants. Eventually they walked the entire state, square mile by square mile, dividing the natural world into small boxes stacked one on the other. I think of the Ojibwe and Sioux still in the state, peoples who had been here "long enough" to know the ground as well as any human ever has known a place, praying in four directions as a way of giving thanks. The surveyors came with a different goal, of course, to know the ground in order to make it easier to divide and sell.

But I bet there were at least some of these surveyors who grew to appreciate knowing this ground more thoroughly than they'd probably known any ground before. Julia nods and points out a blotch of orange on the map near the house I've just rented. "There's a little remnant of original prairie still there, protected by the Park Board," she says. "You should go take a look."

The prairie remnant is a five-minute bike from my house. There is the loud pulse of crickets and the steady buzz of insects, the distance filled with tall yellow-headed flowers through which to wade. There's a scraggled beauty about it, like the messed-up hair and just-awake face of someone you love. The colors are the ochre yellow suns, paste-white buttons, and faded purple lanterns of the wildflowers. And the many shades of brown, the colors of what has been.

To the west, state highway 55 is a constant rush of car tires on pavement, to the east down the slope the river continues the journey it began ten thousand years ago. A single dragonfly patrols, an ambassador from the once-upon-a-time trillions. A bumblebee makes its rounds among the mustard-yellow stars as though nothing has changed. This is all that's left of what was here, a pond where once there was a sea.

When I began this chapter, I was wondering what our relationship with turfgrass says about who we are, the ways it reflects and shapes our worldview. Now, I think part of what it means to be native is knowing what lived where you are before you arrived. Gazing at this remnant of prairie, I can begin to imagine what my home once was like.

It's our obsession with appearances that crowds out what grew here, what and who knew this ground. But in the end, I see opportunity. For most of us, turfgrass is our most direct contact with "natural" ground. If we could see the ground beneath us—in this case, the turfgrass of our lawns—as worthy of being intensely loved, how would that change our lives?

The Sandhills

Who cares about cranes?—and tigers and songbirds and
sparkling streams and hoary ancient forests and traditional
earth peoples clinging to old quiet ways of their language
and culture—or cares enough to defend and protect what
remains of the old world of unbroken and unpolluted nature
on our ever more disrupted mother earth.

— Peter Matthiessen, *The Birds of Heaven* (2001)

By every account, the grasslands of north-central North America were incredibly wild, an ecosystem colored "vivid, as if God had just dyed it" (1905) and bursting with life, with great herds of buffalo, and skies sashed with long flights of passenger pigeons, geese, ducks and cranes. Explorers found grasses "with seed stalks from six to ten feet high, like tall and slender reeds waving in a gentle breeze" (1817) and roots going fifteen feet deep. They described "a vast ocean stretching out until earth and sky seem to meet" (1857), a "prairie-ocean" of beauty each season, "in winter, a dazzling surface of the purest snow; in early summer, a vast expanse of grass and pale pink roses; in autumn too often a wild sea of raging fire" (1872). Before then, in 1835, when Stephen Kearny led a party into northern Iowa, wild strawberries made "the whole track red for miles and stained the horses' hooves and fetlocks," and he and his men rode "stirrup-deep through young bluestem grass and flowers, the hooves muffled in a loamy wealth that had been accruing annual interest for twenty thousand years."

Even as recently as two hundred years ago, tallgrass and shortgrass prairie covered much of North America, from central Canada down through Minnesota and Iowa all the way to the Gulf of Mexico, and east to west from Indiana to Nebraska, Kansas, Oklahoma, the Dakotas. Most of that wild grassland is gone now, some of it paved, much of it tilled—of the tallgrass prairie that once covered 40 percent of the lower forty-eight states, only 2 percent remains. I have come west from Minneapolis to see some of that 2 percent, a road trip with stops first here, near the North Dakota border, then down to Nebraska's sandhills to see a hundred thousand cranes.

My host is Ray Norgaard, a veteran of more than four decades with the Minnesota Department of Natural Resources. We meet in a gas station parking lot just off the interstate where we will leave my car and take his silver Chevy truck into the western Minnesota countryside he knows so well. Dressed in a green flannel shirt and dark green canvas pants, Ray wears a faded gold cap that reads "Ducks. Wetlands. Clean Water." And this is why I have come to spend the day with him, driving straight dirt roads framing endless fields, exploring one of the great bird flyways in the world.

Stretching north from northern Iowa across five US states and three Canadian provinces is the three-hundred-thousand-mile prairie pothole region, so called because of the pockmarks left behind some ten thousand years ago by retreating glaciers—ice that in places was five miles thick. As this ice retreated, it left millions of depressions in the ground, sometimes forty per kilometer, a kind of Swiss-cheese landscape.

And here's the key: while for many of us the word "wetlands" might conjure a pond or swamp, these temporal, or ephemeral, wetlands are wet for only a few days to a few weeks each year. The rest of the time, they look like just another sway in a sea of rolling ground. But in the short period during which they hold water, they come alive with invertebrate life, the aquatic insects that fuel a duck's flight.

At one time some twenty-five million potholes covered this region, and more than a hundred million ducks—maybe far more than a hundred million—filled the sky. Even now, some twenty million ducks pass through the prairie pothole region—pintails, mallards, gadwall, blue-winged teal, shovelers, canvasbacks, and redheads (not to mention birds heading farther north to boreal forest and the Arctic—wigeon, green-winged teal, Canada geese, snow geese). This ground is host to "one of the five greatest animal migrations on the planet."

So says Johann Walker, director of Conservation Programs in North and South Dakota for Ducks Unlimited, a group dedicated to conserving wetlands and ducks. He explains that the pothole wetlands produce high levels of invertebrate biomass, which "is really important to ducks when they're breeding—for egg formation, for maintaining their own body, and for growing ducklings. These wetlands are unique in the world—there's not another prairie pothole region anywhere on the globe."

From a duck's perspective, these shallowest and most ephemeral of wetlands are the best wetlands. But from a farmer's perspective, they are easily drained for the growing of crops. In western Minnesota, most of this drainage took place before the 1930s, as the tallgrass prairie disappeared.

Since that time, North Dakota has lost 72 percent of its native grassland (turned into cropland), South Dakota 64 percent, and Montana more than 55 percent. In the Canadian provinces of Saskatchewan and Manitoba, the losses are even worse, and Alberta has lost half its native grassland. Over the entire prairie pothole region, about 40 to 50 percent of the original wetlands have been drained—though in Iowa and Minnesota that figure is closer to 90 percent. What troubles Ray, though, is not so much what has happened as what is happening: the increased use of new drought-resistant corn varieties combined with high prices for corn and soybeans has led to rising land values and a surge of wetland loss in prairie Canada and grassland loss, especially in the Dakotas, where duck habitat is under serious threat.

"We have been in a wetland crisis for more than half a century," Ray tells me. "On the prairies of Minnesota we've lost ninety percent of our wetlands, and in some counties almost all. In addition to that, we've got a wetland quality issue." In the fall of 2015, the Minnesota Pollution Control Agency released a study showing what he means. Among their findings: on Minnesota prairies only about one in five of the remaining wetlands have good quality. You have to get up into the northeast part of the state before you find good-quality wetlands.

"So," Ray continues, "in this part of the world we're down to the last ten percent or less of our wetland basins, and then fifty percent of those are really of very little benefit. They retain some floodwaters, but from a wildlife standpoint and from an ecological standpoint, they're just bathtubs."

We pull off the road and walk into a depression, a shallow bowl in the middle of a field. Ray tells me that fifty years ago, the ratio of temporary and seasonal basins to permanent basins was 20 to 1. But now, he says, that ratio is only 5 to 1. This depression in which we stand is one of those temporary basins, the kind that hold water only for a few days or weeks. They are the most important for hens for nesting, and critically important for migrating shorebirds looking for food. This is the wild ground that keeps this migratory flyway alive. And, I admit to Ray, if he hadn't stopped the truck and pointed this out to me, I would never have looked over here and known what I was seeing. "Oh, absolutely," he says. "Most people wouldn't."

Back in the truck, we watch as a pheasant brood—a half dozen chicks, little golden-brown, black-striped fur balls—scurry off the road ahead of us. "I bet they're not even three weeks old," he says, laughing.

Ray's laugh reminds me of the animation in Johann Walker's voice when he told me of his first year in the prairie pothole region, and how he would periodically visit duck nest sites to see if they had survived. "And you could pick up the egg. And the little duckling inside had maybe just busted a very small hole in the shell, and you could whistle, just a *peep-peep-peep* at the egg, and the little duckling inside would whistle back at you." He paused. "That was a big moment for me, just to be standing out on all that prairie and to have a vague notion from our sampling hundreds of acres every day of how many hundreds of nests were really out there on the ground, with how many eggs in them, and here I was holding one of those in my hand, watching that little life emerge."

The original grasslands in all their glory are something we will never see. "The most decimated ecosystem in continental North America," the prairie is sometimes called. Gone are the grizzly and the wolf, and the populations of nearly every species—bird, animal, plant, and even insect—are severely reduced. In the past decade, as the monarch butterfly began to disappear, it marked the latest species to decline.

But still there is a wildness in the ground that brings the prairie potholes alive each spring, for days or weeks or months at a time. Dry beds in the landscape, shapes to which most of us would be blind, are the shallow basins in which so many lives will form. The ducks that visit our city skies, that flash past in fast Vs over suburban lakes and ponds—those ducks are made of this ground. This is the fountain of energy that Leopold spoke of, rising from soil into sustenance into brown and gray-white bodies, cinnamon and emerald heads. This is that fountain rising into beaks and feet, eyes and wings.

"What really drives the system is invisible," Ray says, turning the Chevy back toward the interstate. Yes, but sometimes, it seems, the evidence of that invisible life is everywhere for us to recognize. Just a day's drive south of here, one of the world's greatest remaining displays of wildlife still settles from the sky to share the ground on which we stand.

It is from the oldest of worlds, an ancient sound and sight still available to us today, the wild birds standing four feet tall, their wings wrapped like shawls. The sandhill crane, which at twenty million years and counting is the oldest living bird species on Earth, comes to us essentially unchanged through time, essentially the same bird that has always been here.

In predawn darkness, I rise with excitement and come east on I-80 before turning off to follow red taillights, a long line into the parking area, a blustery morning with last night's Milky Way hanging washed in the southern sky. And then, a crimson waning gibbous moon near the horizon, I hear them in the distance, the unmistakable call that had drawn me from sleep to come bundled in wool and down to join a hundred others at the edge of Nebraska's Platte River. After introductory remarks from the Rowe Sanctuary folks, we walk without words to our lookout points, wondering how many we will see, whether we will luck out or miss. The blinds are wooden structures like mobile homes, with paneless window cuts on the river side. We slide through the door and take our positions, to wait and watch and listen.

Since reaching Kearny yesterday and seeing my first cranes, I have remembered how much of their appeal comes from their call, a sort of purring, trumpeting trill caused by the unique shape of their larynx. At this point this morning the calls are grounded, as several thousand sandhill cranes stand on sandbars in the river before us. The volume of calls and the light in the sky seem to grow in sync, and then on some undetectable sign the cranes lift from their sandbar roosts in the shallow river and leap into the orange light that just now comes over the horizon, a sudden cacophony lifting into the dawn air, golden light on bellies and wings, first one section of the great flock then another, leaving the place where they spent the night, and a sort of euphoria ripples through the blinds as the gathered humans smile and want to cheer but know to stay quiet. After the bulk are gone you still hear from straggler cranes echoes of what was here.

"Whichever road I follow, I walk in the land of many gods," writes Linda Hogan. My senses still humming from having been on the ground that launches these descendants of ancient birds, I think of her words. "Suddenly all my ancestors are behind me. Be still, they say. Watch and listen. You are the result of the love of thousands."

*   *   *

I first saw sandhill cranes in a New Mexico November twenty years ago. Hiking with friends at Bandelier National Monument, we watched as one drifting flight after another of these majestic birds floated over in the deep blue sky between the canyon walls. They were flying south following the Rio Grande, many making their way down to the Bosque del Apache National Wildlife Refuge seventy-five miles south of Albuquerque. Since then I have seen cranes many times at the Bosque, traveling various dawns to watch the great rafts of birds leaving the refuge before sunrise, and various dusks to watch them return after a day of feeding in surrounding fields. But until this visit I have never had a chance to see the annual migration of between four hundred thousand and six hundred thousand sandhills coming through central Nebraska on their way north. It's a migration that has been going on seemingly forever, and that now relies almost entirely on our treatment of the ground.

On a map of North America, the crane migration routes flow south from the top left, from Alaska down through Canada, and from the top right from the Canadian Arctic, the two streams bending together like an hourglass with the narrowest waist over this part of Nebraska, then spreading out southwest to New Mexico, Texas, Mexico, and southeast to Florida and Cuba. They didn't all used to come through on such a narrow band along the Platte. But due to human development on traditional resting, feeding, and mating areas, the bottleneck has formed. Agriculture along the Platte has increased crane numbers by offering a landing area with abundant food. Since the 1950s, the mechanical corn picker has provided the cranes with an almost unlimited high-energy food source, the corn left over after harvesting. That's why if you travel to the Platte in the spring you will see groups of cranes spending their day picking through the fields. Typically, the cranes stay for several weeks, adding about 20 percent to their body weight before continuing north. In their long migrations, the cranes rely on stopovers like this one on the Platte to store enough fat to reproduce successfully.

But their future is anything but secure. Corn farming is becoming ever more efficient, and the "beans" part of the "corn and beans" recipe for American farmland has reached these plains as well. The area around the Platte that once was purely corn is now about half soybeans, and as wildlife biologist Gary Krapu tells me, "Soybeans are worthless to cranes." Krapu, who recently retired from the US Geological Survey after a career working with cranes, adds competition for water in the cranes' wintering grounds and hunting along their migration routes as additional threats to their survival. And looming over everything is the threat posed by a changing climate.

"Cranes have been here a very long time," he says, "and they've seen huge changes in the climate on the North American continent. But they didn't occur overnight, and they were of a different nature than what we're experiencing now with human impacts." Already, of the world's fifteen crane species, twelve are listed as endangered. And while the sandhills are the most numerous species, a simple change in agricultural policy—such as favoring soybeans over corn—could, one biologist told me, "send the cranes into a tailspin."

It's hard to believe that a species as old as this bird could now be so fragile. From the first moment I saw them, these creatures with their six-foot wingspan have taken my breath away. The more I learn about them, the more I am amazed. For example, Krapu tells me that pairs will return to the same breeding ground, even the same nesting site, to which they migrated the previous year. Returning, that is, from many thousands of miles away.

But all around the world, centuries-old long-range bird migrations like that of the cranes are being slowed or stopped by our use and abuse of the ground. Recent headlines include "Migratory 'Flyways' Decimated by Human Expansion," and "Habitat Loss Seen as Rising Threat to World's Migratory Birds." As I had seen in the prairie pothole region, migratory birds need healthy ecosystems along their path to serve as rest stops on their often transcontinental trips, yet at least half the world's wetlands have been lost during the past century. In a recent study published in *Science*, researchers examined the "migratory routes, stopover locations, breeding grounds, and winter locations of 1,451 migratory species" and found that 91 percent traveled over ground that is not protected. Because migratory species by definition move from one area to another and rely on a series of landing sites to provide food and rest, explains Richard Fuller, an Australian conservation scientist, "if even a single link in this chain of sites is lost for a species, it could lead to major declines or even its extinction." Another of the study's authors gave the example of the bar-tailed godwit, which migrates from the Arctic to Australia, stopping along the way at mudflats in China, North Korea, and South Korea. "Many of these critical sites have been lost to land reclamation owing to urban, industrial, and agricultural expansion," says conservation scientist Claire Runge, "and the species is undergoing a rapid decline."

I remember being in the blind during the morning, looking toward where the cranes had just left the river, realizing that they had been out there all night. I mean, it's not as though they all sail into a warm, safe airplane hangar every evening. No, every evening they sail instead to the sandbars from the fields where they spent the day. They then stand there through the night, waiting for dawn. The whole time I'm in the hotel sleeping, waiting to rise to witness them, they are out here. And the rest of the year while I'm doing whatever I do, they are out here—or in other wild areas, dodging bullets, finding food, waiting for currents in the wind, returning to the very ground they knew the year before, pulled by some old yearning at which we can only guess. Of course our choices affect them. Of course our bullets knock them from flight. Of course if we take away the grounds on which they depend, their numbers will dwindle to a whisper of what they are now. But otherwise, they exist as they have always existed.

When they fly their flocks often swirl and seem without direction, a gathering stream. Then from a distance their long flights look like pieces of black thread laid out against the purple dusk. At twenty years or more they are long-lived birds, and they will sometimes fly five hundred miles each day. For me, their migrations mean something similar to seeing the stars. They are evidence of something that includes us, but is bigger than us, of a force that goes on around us and without us. But with our building and paving we have halted so many wild migrations, from large to small, the movements that animals and birds and fish and reptiles and insects (monarch butterflies again come to mind) have followed forever, circulating through the world as the blood through our heart and limbs, the breath through our lungs, the spirit that gives us life.

I think about a conversation I had with Jeb Barzan, director of field ecology for the International Crane Foundation, about the value of the ground for a bird that migrates so many miles each year, spending so much time in the air.

"All fifteen species of cranes nest on the ground, and they can nest nowhere else," he said. "Although these birds can fly, the first ten to twelve weeks after they hatch they stay on the ground, and their survival depends on their ability to be safe there. It's a very vulnerable time, and they live or die based on that strategy." The cranes' ultimate survival is going to be dependent upon how people use those grounds, Barzan explained.

And if we care about that survival, we cannot rely only on "protected" grounds. In North America, for example, only about 25 percent of the land is owned by government; 75 percent is privately owned, and most of that is in agriculture. As Barzan told me, "It doesn't take a rocket scientist to figure out that if we're going to provide for biological diversity, of which cranes are a charismatic representative, we have to focus on what happens on those private lands every bit as much as what happens on those public lands."

Two recent studies raise troubling questions about our ability to do that. In the first, researchers found that development and population density around our national parks has greatly outpaced growth elsewhere. While the percentage of population density increase in the United States since 1940 has been 113 percent, around national parks it has been nearly double that, at 224 percent. And while the percentage increase in population densities around specific parks varied (for example, 210 percent around Glacier and 246 percent around Yellowstone to more than 3,000 percent around Mojave National Preserve), every park in the national system saw human development increase around its borders. The second study focused on the National Wildlife Refuge System, a nationwide system of nearly 500 units. In results similar to those of the national parks study, researchers found that the amount and density of human development around refuges far outpaced national averages.

What this means is that the wild grounds we have designated as protected are slowly—or not so slowly—being encircled by human development, becoming islands of wildness in an otherwise tame sea. This matters for several reasons. For one thing, animals don't recognize artificial boundaries between public (protected) and private land. As our populations encircle the areas used by wildlife, there are bound to be more human–wildlife confrontations. Other problems for wildlife include the introduction of invasive species, "subsidized predators" such as feral cats, chemical pollution from pesticides and fertilizers, and the disruption of historical migratory routes. This last consequence is probably the thing that worries scientists most. If wildlife cannot move from one area to another to mate or feed, their ability to survive will diminish in a world rapidly being changed by a disrupted climate.

In the evening I park and walk a long way in constant wind, following the sound, drawn toward them, wanting to be close. I continue walking past the bridge where the photographers and tourists wait, past someone's house—a giant swing in the yard, they must have children—and think *What on earth must it be like to live and grow with a visit from five hundred thousand cranes every year?* I turn down a long dirt road. I feel the wind, see a yellow setting sun in the wheat, feel single raindrops splat my cap's bill, my hands, my feet. I follow the ancient sound of cranes returning at dusk, the great flights descending, coming in after the day of gathering energy from the cornfields, this wild, ancient ritual repeated here before me in a world gone so crazy and broken and out of control. All these birds that have moved across the continent longer than any other, that predate anything we have done, are here still alive.

I stop in the road. To my left are the gathering flocks, to my right a field of wheat. It is like being up close to something back in time. Like a secret still going on. What are we if not a creation of days and experiences, of moments we have stopped to notice the world—the faces of family before us, the beauty of distant mountains, the ground at our feet. I am made from times like these, when I follow whatever voice I heard—who knows where it comes from?—and walk out alone along dirt roads to find a scene like this.

I turn into the field opposite the cranes and lie down on the earth, my surrounding horizons the swaying stalks and tassels, my sky dark blue-gray billows lit yellow-gold by a declining sun. Again and again the long flights pass over me, some higher, some only a few dozen feet above, wings braced for slowing, long gray feathers together in perfect ancient design. Sticklike legs and trident feet extending back, hanging down. Crimson-red crowns, black-fringed tips to blue-gray wings, shaded with gold light, the colors of thunderclouds.

As red-winged blackbirds sing, setting sunlight shines on the cranes' gray bellies, on the plains, as though the same golden energy that fills their wings fills the ground on which their feet will land. This ground feels akin to that of the West—the dry air, the cottonwoods, the big sky—even though it's of the plains. What has passed beneath me? A thousand thousand years of animals moving, eating, mating, dying—the bones below us we never will know. This outermost surface of the earth all the while becoming the richest soil.

Single drops of rain, and the steady calls of cranes. In his essay "Thinking Like a Mountain," Aldo Leopold described the wolf's howl as "a wild, defiant sound." The crane's call is another of the world's wild, defiant sounds. Gathering for the night, it's as though they're calling to friends and family stories of the day, all the amazing things they saw and found. And what a life, journeying north and south with the seasons, and all the little cranes, colts that will be born into this threatened world, what world will they come into? But they will come. I have faith in that.

Booms of thunder welling in the east and south chase me back onto the road. The dark horizon backdrop is filling with lightning stems and blooms. I hate to leave, but great bolts brand the sky. It's like walking away from a rushing river. Will they sound through the storm? Or do they quiet, listening for creeping coyotes?

HELL AND SACRED

The grounds beneath us, which will we choose?

Appalachia

It is a curious situation that the sea, from which life first arose, should now be threatened by the activities of one form of that life. But the sea, though changed in a sinister way, will continue to exist; the threat is rather to life itself.
    — Rachel Carson, *The Sea Around Us* (1951)

Like almost everyone, I had heard about it. Even Merriam-Webster had finally heard, adding the word in 2014, along with "hashtag," "unfriend," and "selfie." But me, I live in Minneapolis, and there is no oil and gas exploration here, no "fracking." I grew up surrounded by lush green leaves and blue lakes we could swim, with loons in summer and the Mississippi flowing through. This is my home—educated, liberal, hipster beards and bicycles. We would never let fracking happen here. But we know it's happening somewhere—the oil rigs, the man camps, the noise and dust and light. We may even know about the pollution, the spills, the derailed night-trains billowing flames so hot the local fire guys stand at a distance and watch. These are rumors of war, news from a faraway land.

So I travel to southeastern Ohio near the West Virginia border, Appalachia. I go to where a local woman tells me, "They all think we're just retards down here." And I feel I am visiting a country where a foreign force from a land of white pickups and only men, stained and unshaven, has come with its heavy machinery to do as it pleases. I think, *This is an America I don't know well, that maybe most of us don't.* I think, *This reminds me of hell.*

That local woman's name is Carrie. We meet at McDonald's, she and her sister Sara, in their fifties, wearing tight jeans and straight hair hanging past belt buckles, and Carrie's husband, Jeff, he of the three missing teeth and one silver one. There's a sense of going undercover or behind enemy lines as we meet and plan our drive. Or maybe that's just me, standing quietly as Ted, my host, helps figure out the route. I feel out of place. Their accents are Appalachian, and I have to listen hard to understand. I smile and shake hands and give thanks that Ted is here to translate. Ted, who works in Ohio for an organization dedicated to keeping track of fracking operations in the state, has introduced me—I would never have met them otherwise. They're poor, like most people around here, and they feel betrayed by a fracking industry that promised riches.

"This community that I live in is mostly uneducated," says Carrie, sitting in the backseat between Ted and me. "In southeastern Ohio, you're going to find fewer college graduates. You are going to find people that work in factories, that lost their jobs and haven't had a job in three years—most of them longer. They go to church every Sunday. They read their Bibles. They believe in God. They believe in paying their taxes and going by the rules. And they think everybody does that."

In other words, a receptive audience for a fracking industry selling easy money?

"The one thing I really would like everyone to know is just

how they do this, when they come in," Carrie says. "When this all started in 2011, the first thing these companies did is call a meeting in the basement of the local church—and that's the start of the lie, right there. They do that in every community. Then they tell everybody they're going to be rich. They may not say those words, but everybody is high-fiving. They haven't had jobs."

Carrie and Sara and Jeff tell us the industry goal is to entice as many landowners as possible to sign leases allowing drilling on their property. If landowners resist, the industry shifts gears, pitting neighbor against neighbor and using a process of "unitization" that allows them to tap the oil and gas lying under the properties owned by those who don't want drilling on their land.

"Here's the thing that nobody really knows," Carrie says, "just what a predatory industry this is. With unitization, they'll tell you like this, 'You can either sign with us, or you are going to be force-pooled. That's your choice. We are going to take it.'"

"And this is what they won't tell you," Sara says, turning from the front seat. "Maybe later on down the road, because they've got the rights, they got you in a unit now, they can do this, and this, and this, and this. Anything they want."

"They can kill your mother and cut her head off if they want to," says Carrie, "because that's unitization."

The word "fracking" is short for hydraulic fracturing, a process of extracting gas and oil from the ground. The process has been around since the 1940s, but recent advances in technology have led to an explosion of fracking activity across the United States and in many other parts of the world. Some Americans think this the greatest thing ever. They point to money and jobs; they wonder how anyone could be against it. Other people, like the commissioner of health for New York State who said he wouldn't want his family to live in a community where fracking was taking place, and the governor of New York, who banned fracking in his state, haven't thought it's so great. They point to the potential long-term health costs. "We cannot afford to make a mistake," the commissioner said. "The [significant public health] risks are too great. In fact, they are not even fully known."

So how does fracking actually work? At Halliburton.com, under "Hydraulic Fracking 101," here's what they say: "Well, it starts with a good bit of water and a lot of sand. Mix those two together, apply a couple thousand pounds of pressure, and introduce them to a reservoir several thousand feet below, often with the help of a small percentage of additives that aid in delivering that solution down the hatch." From there, the blasted water (plus additives) creates tiny fissures in the rock that allow previously inaccessible oil and gas to rush to the well, and voilà, we have a producing well. This process takes, Halliburton says, "3 to 10 days to complete."

Next, "It's time for the operator to remove the water, clearing the way for the newly stimulated well to produce energy for the next 20, 30, 40, even 50 years. The trucks, the pumps, the equipment, and the traffic that were needed to do the job—they're long gone. The operator typically leaves a production valve and collection equipment behind. The rest of the site is remediated, often within 120 days."

The folksy language ("a good bit of water," "down the hatch") makes these guys sound superfriendly. Heck, they're only using "a good bit" of water and just a few "additives." The trucks and trucks and more trucks are soon "long gone," and everything is pretty much back to normal in a few months. Except now we have more energy than we ever dreamed possible. What's not to like?

Of course, Halliburton has an interest in making fracking sound simple and clean. They invented hydraulic fracturing in the 1940s and are one of the world's largest suppliers of products and services to the industry. In the United States the name Halliburton is well known because Dick Cheney, vice president in the George W. Bush administration, was at one time its chief executive. The name Halliburton is also familiar because of something called the "Halliburton loophole," which Cheney ushered into law in 2005. This loophole exempts the fracking industry from key provisions of the Clean Water Act, the Clean Air Act, and the Safe Drinking Water Act, essentially stripping the Environmental Protection Agency of its ability to regulate the process. A key aspect of this loophole is that it allows the fracking companies to keep the identity of their "additives" a secret. Their argument is that these additives are proprietary—like McDonald's "special sauce"—and that having to reveal them would hurt their competitiveness. Because fracking is basically safe, their argument goes, why would it need regulation in the first place?

In fact, in this very poor part of Appalachian Ohio, the narrow rural roads are full of trucks, flatbeds hauling heavy machinery, white pickups stuffed with men in helmetlike hardhats, and ubiquitous tankers labeled "Brine." Most of these trucks sport plates from other states, Texas and Oklahoma especially. Around every corner there's another truck of some kind, the gas and oil industry on wheels. Trucks in newly scraped-bare ground lots. Trucks pulled to the sides of roads. Nowhere in the places fracking is happening were the roads built for so many trucks. And no one who lives where fracking happens has ever seen so many trucks at once, sometimes dozens an hour. No one has ever heard or smelled anything like the noise, exhaust, and fumes. The trucks Halliburton promised would be "long gone"? They're down the road at the next new well.

The Halliburton description of fracking tries to reassure us, like, "Heck, you'll hardly notice that the industry has come to town." But in real life it looks like the fracking industry has taken over the town. It certainly looks to have taken over the ground. In fact, the ground has the look of recent battle, torn up and scraped bare, all yet to heal. Long corridors slashed through woods for new pipelines, fresh wounds, fragmentation—forests and fields broken apart. Compressor stations, injection wells, pump jacks, well pads—all of the infrastructure that fracking demands, all of this so new to Ohio, the conquering army recently arrived.

Contrary to the industry's implications, the impact of fracking on the ground is tremendous. In a recent study designed to gauge the ecosystem services lost to oil and gas development—such as the ability to grow food, store carbon, support wildlife—scientists found that nearly eight million acres had been stripped of vegetation, the equivalent of three Yellowstone National Parks consumed. "For all intents and purposes, these are parking lots," says Brady Allred, an assistant professor of rangeland ecology at the University of Montana. Compared with the size of the United States, this acreage might not seem so much. But we are almost certainly nowhere near the end of the fracking boom; fracking will impact a lot more ground in the next ten, fifteen, twenty years. And because of the roads and other infrastructure needed to support it, each well pad can have an ecological footprint of more than thirty acres. For the past ten years, the industry has been drilling fifty thousand wells a year, a trend that many would like to see continue. "Whenever we tell people, they think we're crazy," says Allred. "Nobody has any idea of the magnitude of this."

Carrie, Jeff, Sara, Ted, and I take the highway north, then exit and follow back roads until we coast to rest on a hilltop. What we see shocks me. The next hilltop over, maybe one hundred yards away, has been sheared, flattened, the bare brown ground visible where the natural ground cover has been peeled away. And on this bald hill, scraped into the ground like a new tattoo, bandages just removed, an enormous compressor station rumbles, with several tall pipes burning orange-red flares, like pedestals topped with columned bonfires, their billowing exhaust blackening the bellies of spring clouds. The sparkling new home for oil and gas looks like a warehouse thrown up overnight; the old hills on which all this is built seem stunned. I stand transfixed by the towering flames, my mind saying *Something's on fire*, and indeed something is: natural gas deemed less valuable than the oil for which we hunger. And so we burn this gas day and night to get it out of the way, burn the bodies of ancient plants and animals, sending them skyward in the form of smoke and flames.

The constant churn and whine of industrial noise assaults our ears. The sounds of nature have been disappeared—no birdsongs, crickets, or breeze, just that incessant decibel blur. I have seen industry videos that tout the "noise canceling walls" built around compressor stations, but I never see this in Ohio, in country that until fracking came was quiet. Fences circle most compressor stations, with manned gates through which trucks exit and return. Watching them feels like watching a jail from a distance—somewhere you can't go, somewhere you don't want to go. Or like watching a foreign military base on your soil, the occupiers hidden somewhere inside.

Natural gas pipelines need compressor stations at regular intervals to keep the fracked gas moving. At this first compressor station, at the base of the scalped hill, sits Jeff's grandmother's house, small and white. "We was playing here as kids," he says, looking at the compressor station "This was her and my grandpa's land." The fracking folks wanted this hill, and kept after Jeff's grandmother—who did not want to move no matter how much they offered—until she finally gave in.

Imagine having a factory/prison/military base with flaming spires move in behind wherever you live. Most of us can't. But in southeastern Ohio I see the ground we are willing to sacrifice in order to cheaply fuel our lives. Far too often these sacrificed areas have been in foreign lands, some country we have never heard of and to which we will never travel. Or these sacrificed areas were long ago lost, ravaged before we were born. What's shocking is when you visit these newly cleared areas and try to make sense of the fact that you're in present-day America. You look around in disbelief, wondering whether you aren't really somewhere else.

Where is—or *what* is—hell? Do we still imagine it as an actual place, an underground lair of torment? Even the Catholic Church now dismisses this idea. As Pope John Paul II explained in 1999, "The images of hell that Sacred Scripture presents to us must be correctly interpreted. They show the complete frustration and emptiness of life without God. Rather than a place, hell indicates the state of those who freely and definitively separate themselves from God, the source of all life and joy." Yet the popular images of hell live on.

And they don't really come from the Bible. According to someone who took the time to count these things, there are only fifteen mentions of hell in the Bible, compared to 622 mentions of heaven. The idea of hell as a place of flames and torment comes especially from Dante's *Inferno*, the first book of his *Divine Comedy*, written between 1308 and 1321. Dante divides his hell into nine levels descending through the bowels of the earth and organized according to a justice system of *contrapasso* ("suffering the opposite"), in which souls are punished according to the nature and degree of their sins. You might spend eternity being blown about forever in pitch darkness, or getting torn apart by three-headed dogs, or endlessly rolling boulders; it depends: "Sowers of discord are cut to pieces, flatterers swim in a stream of excrement, and a traitor spends eternity having his head eaten by the man he betrayed." Dante punishes worse sins with burning in tombs, boiling in blood, dunking in boiling tar, and ripping apart with hooks. He describes the devil (with his awesome title "the Emperor of the Universe of Pain") as standing frozen up to the breastbone in a great lake of ice, "his leathery bat-like wings beating furiously."

Medieval Christianity did not have a monopoly on colorful descriptions of hell, especially the idea that the worse the offense, the worse the punishment. Hinduism has twenty-eight main hells (with possibly hundreds or even thousands of sub-hells), and Burmese Buddhism more than forty thousand, "one for each particular sin, including nosiness, chicken-selling, and eating sweets with rice." The Egyptian hell "was a fiery pit in which people were not merely burned, but hacked into small pieces by 'the Commander of Fire' and 'the Eater of

Entrails'; and both the Hindu and Buddhist hells put sinners in cauldrons, or on spits basted by many devils, the better to feel the heat." Not to be outdone, traditional Chinese culture consists of several levels of hell, including the Chamber of Tongue Ripping, the Pool of Blood, the Town of Suicide, the Mountain of Flames, the Chamber of Saw, and the apparently closely related Chamber of Dismemberment. The world beneath our feet was quite a busy place in days gone by. Legends told of travelers discovering holes in the ground where smoke rose from hell-fires, and even in our own day an urban legend persists of engineers halting their drilling of the world's deepest hole (some nine miles down) when apparently they heard human screams and thought they had literally drilled into hell.

English poet John Milton took Dante's cue and in *Paradise Lost* further reinforced our notion of hell as underground torment. "As far removed from God and light of Heaven . . . Oh how unlike the place from whence they fell!" For Milton, hell was less flames and pitchforks and more darkness: "A dungeon horrible, on all sides round / As one great furnace flamed; yet from those flames / No light, but rather darkness visible." Moving along through the ages, in Percy Shelley's "Peter Bell the Third," hell is imagined as "a city much like London — / A populous and a smoky city; / There are all sorts of people undone, / And there is little or no fun done; / Small justice shown, and still less pity."

These days, though, hell has lost its precise location. As the pope says of purgatory, "The term does not indicate a place, but a condition of existence." Of course, this can be bad enough. I think of William Styron's powerful "memoir of madness," *Darkness Visible*, the title taken from Milton's phrase. It seems far more useful to understand hell this way—as a situation rather than a setting, and especially as mental torment—rather than as a place deep in the ground.

Maybe no one in history has depicted hell more vividly than the early-sixteenth-century artist Hieronymus Bosch. During his career, Bosch painted images of hell several times, and never more evocatively than in *The Garden of Earthly Delights*. This painting is actually a series of three paintings on three panels, a triptych that tells a story from left to right. On the left panel is the Garden of Eden, in the center (and largest) panel is the beauty of life on earth, and the right panel is Bosch's depiction of hell.

From left to right, humans move from paradise to damnation. We take a world of greenery and wildlife, overpopulate it with ourselves, and end up with a world of sterility and suffering. It's worth noting that in the first two panels Bosch portrays the ground as green and alive, while in his depiction of hell he paints that same ground a barren tan. The skies, the waters, the land—all have been emptied of animal life—in fact the only animals depicted are monsters devouring humans (a giant rabbit carrying a decapitated human corpse, for example). Where once humans, plants, and animals shared a living world of green and blue, now war and destruction reign, cities burn against a black background, and human life is reduced to torture and grief.

Bosch's painting reminds me of the dead world in Cormac McCarthy's novel *The Road*. Hailed as the most important "environmental" book since Rachel Carson's *Silent Spring* fifty-some years before, *The Road* is set in a "barren, silent, godless" world, a "cauterized terrain" where "all was burnt to ash," an "ashen scabland" full of dead trees, no sound, no movement. Like Bosch's painting, McCarthy's novel asks, In which direction are we headed?

*　　*　　*

For all our group sees above ground, the occupation is at least as widespread under the surface. We stop at our first injection well. Four enormous drums, industrial barrels twenty-five feet tall, packed together like a giant's stash of beer. Signs warn us away—private property, etc.—but there are no flames, no noise, and hardly anyone's around. This injection well and those many others that I'll see don't seem nearly as diabolical as the compressor stations.

In fact, they may be more so. In the Halliburton description of the fracking process, once the oil and gas are flowing, "it's time for the operator to remove the water" that was used earlier. It isn't just "a good bit of water," though; there are also some "additives." These additives are chemicals, many of them toxic. When the operator removes the "water," he's bringing up toxic waste. What chemicals are in this waste? No one knows—no one like you or me, that is. The companies know, but because of the Halliburton loophole, they don't have to tell anyone. And even though everyone knows the companies are using chemical additives along with their "good bit of water," the waste is considered nonhazardous.

How could this be? In 1988, the energy industry wangled a critical change in the federal government's legal definition of waste: all material resulting from the oil and gas drilling process is officially considered nonhazardous, regardless of its content or toxicity. That's why the tanker trucks clogging the narrow roads of fracking land are labeled "Residual Waste," or simply "Brine," a word that has me thinking of Thanksgiving turkey.

When the operator removes this mystery waste, the contents of which no one but the company knows, it has to go somewhere. It's sloshed into tanker trucks and hauled on narrow roads to the nearest injection well, where it's injected into an old well—back into the ground—and sealed. The fracking industry says not to worry, that it doesn't really matter what's in the waste because it's safely disposed of, so far below the water table, supposedly so far away from where it could do any damage.

This isn't happening only in Ohio. Nationwide each year, American drillers pump more than six hundred billion gallons of fracking waste back into the ground. All told, more than 680,000 injection wells dot the country, holding thirty trillion gallons of toxic waste. Why? While fracking waste can be recycled or processed at wastewater treatment facilities, to do so is far more expensive than simply pumping it into the ground. Also, putting it back underground removes it from sight. Imagine the complaints if companies were made to create enormous swimming-pool-like containers, or huge rain-gauge-like structures, out in the open where we all could see.

So while companies could never legally dump toxic waste directly into rivers or spew it onto soil, they are allowed to pump it deep underground. Until recently, scientists believed this method of storage to be safe. But that thinking is beginning to change. "Are we heading down a path we might regret in the future?" says Anthony Ingraffea, a Cornell University professor and an outspoken critic of claims that injection wells don't leak. "Yes." The hundreds of thousands of holes being punched in the ground are changing Earth's geology, adding human-made fractures that allow wastewater to flow more freely. "There's no certainty at all in any of this, and whoever tells you the opposite is not telling you the truth," according to Stefan Finsterle of Lawrence Berkeley National Laboratory. "You have changed the system of pressure and temperature and fracturing, so you don't know how it will behave." In short, critics say, the science that permitted the deep wells of fracking waste hasn't kept up with the current boom. And because all this is happening far below ground, by the time any damage is discovered, it could be too late.

The "too late" scenario that worries scientists most is contaminated groundwater, because once contaminated, groundwater is virtually impossible to clean. In the United States, we rely on groundwater in a big way. Collectively more than 50 percent of Americans—currently some 150 million people—rely on groundwater for their drinking water (95 percent of households in rural areas, 35 percent in urban areas). And when it comes to agriculture, the numbers are even more staggering: more than 80 percent of our nation's agricultural water needs are met by groundwater.

Here too the oil and gas industry claims there is no danger of contamination from injection wells (or from surface spills), but many aren't so sure. "In ten to one hundred years we are going to find out that most of our groundwater is polluted," says Mario Salazar, an engineer who worked for twenty-five years as a technical expert with the EPA's underground injection program in Washington. "A lot of people are going to get sick, and a lot of people may die."

I doubt that many of us truly believe that deep in the fiery bowels of Earth countless humans endlessly toil amid fiery furnaces, whipped to work by goat-horned devils. But I wonder if whether we actually believe that hell is "down" (or that heaven is "up") almost doesn't matter, because the metaphor is so ingrained in American culture. That hell is in the ground tells us Earth can be used however we like; that heaven is in the sky tells us paradise is somewhere else. As the Mormon pamphlet I once received put it, "The Earth is not our home."

Barring a nuclear holocaust—or a world in which giant rabbits devour us for supper—a far more likely scenario for "hell on earth" has to do with altering the planet through our way of life. The effects of climate change are becoming apparent around the globe, and many say of known fossil fuel reserves that we must "leave it in the ground" to avoid catastrophe. But energy companies are spending billions every year ($600 billion in 2014 alone) to find more, and governments around the world are spending another $600 billion annually to subsidize fossil fuel use. Biologist and author Sandra Steingraber calls the issue of whether to continue fracking for oil and gas "the question of our time." But in Appalachia as elsewhere, we continue to pull dead ancient life from the ground, sending it around the world, setting it aflame.

The result? The UN's Intergovernmental Panel on Climate Change warns, "Continued emission of greenhouse gases will cause further warming and long-lasting changes in all components of the climate system, increasing the likelihood of severe, pervasive and irreversible impacts for people and ecosystems." It describes a world threatened by "food shortages, refugee crises, the flooding of major cities and entire island nations, mass extinction of plants and animals, and a climate so drastically altered it might become dangerous for people to work or play outside during the hottest times of the year." It's worth keeping in mind that the IPCC is generally seen as conservative in its predictions.

Wanting to share a story about where we might be headed, two scientists in 2014 wrote a small, remarkable book. In *The*

Collapse of Western Civilization, Naomi Oreskes and Erik M. Conway took existing research on climate disruption and put it into the "science fiction" narrative of a Chinese scientist looking back from three hundred years in the future. If you want a detailed description of hell on Earth, look no further.

"Then, in the Northern Hemisphere summer of 2041," their Chinese scientist writes, "unprecedented heat waves scorched the planet, destroying food crops around the globe. Panic ensued, with food riots in virtually every major city. Mass migration of undernourished and dehydrated individuals, coupled with explosive increases in insect populations, led to widespread outbreaks" of various diseases, and the destruction of "huge swaths of forests in Canada, Indonesia, and Brazil." Social order begins to break down in the 2050s in Africa, Asia, and Europe, and as the drought destroys the mid-American breadbasket, the US government declares martial law and begins talks with Canada on what eventually becomes the United States of North America in order to share resources and facilitate movement of citizens from flooding coastal cities. Things continue to deteriorate from there, and while human society survives, it doesn't look much like the world as we know it today.

Powerful as this story is, what happens to the rest of creation—the animals, birds, fish, reptiles, insects, and so on—is left to our imagination. In fact, scientists now fear a "sixth extinction," in which upwards of two-thirds of existing species could disappear during this century. It's an almost unimaginable situation, certainly one possible definition of a world damned. The French philosopher Jean-Paul Sartre famously said that hell is other people, but I might amend that to hell is *only* other people, with no other creatures around. I think Bosch knew this—his Garden of Eden and his larger middle panel depict a world filled with animals. His third panel, of hell, has far fewer animals, and those present have morphed into monsters.

Already, one new scientific study after the next tells another story of world biodiversity in decline, previously bountiful populations of wild animals reduced—the common birds of Europe by a third since 1950, Africa's lions by 75 percent since 1900, the world's insects by nearly half, and on and on. Here is the threat "to life itself" that Rachel Carson warned us of more than six decades ago.

Whenever someone laughs off concern about "the environment," I wonder whether they have any idea of what's happening.

Thinking of hell, of where we might be headed, I think of someone who—though from a very different era—definitely knew what was happening around him. The English poet John Clare was a man whose close observation of nature spawned some of the richest poetry in Western history—and probably also drove him mad. Clare spent his childhood and youth exploring the rural English landscape about an hour's train ride from modern London, living most of his life within a day's walk of the house where he was born. His early poetry, writes scholar Geoffrey Summerfield, "is primarily a celebration and affirmation of life: of human life and of all forms of natural life—of animals, birds, insects; of the dawn and dusk and of the seasons; of the soil and of weather; of trees, rivers, sunlight and cloud." His firsthand experience in and knowledge of nature sparked and fed his affection for his home, and his early poems are full of this affection.

In "The Nightingale's Nest," Clare wrote of quietly moving through woods, "for fear / The noise might drive her from her home of love." The nightingale will nest, Clare knew, on the ground, but the nest is so well camouflaged that to find one he admits he has "searched about / For hours in vain." He knew this bird's nest ("velvet moss within / And little scraps of grass, and, scant and spare, / What scarcely seem materials, down and hair"), and he knew that his presence "doth retard / Her joys, and doubt turns every rapture chill." And so, in order to not disturb her, "here we'll leave them, still unknown to wrong, / As the old woodland's legacy of song."

In 1809, enclosure came to his town of Helpston. Enclosure meant the closing of what had been common grounds into fenced private areas, a move that benefited wealthy landowners while forcing the local working-class poor from land they'd long known. In an instant, Clare was separated from the grounds he'd known intimately. His world was, says Summerfield, "despoiled often beyond recognition," and his poems describe the agony of losing animals, open spaces, and individual trees that he'd seen as friends. In "The Fallen Elm," Clare writes of an "old elm" that had provided shelter to humans and birds alike, but now the newly emboldened landowner, "With axe at root he felled thee to the ground." Clare mourned too the more intensive agricultural methods that came with enclosure, as "the rage of the blundering plough" tore through the ground he loved, and "each little tyrant with his little sign / Shows where man claims earth glows no more divine."

The effect on Clare's health was devastating. A growing number of scholars acknowledge that "the loss of almost all he knew and loved," as the English journalist George Monbiot wrote in 2012, was at least a significant contributing factor to his decline. It's not hard to believe. To read Clare's poetry is to experience the natural world with someone alive to its intricacy and mystery. His is a poetry of biophilia, of human love for the natural world. Long before any such word existed, Clare described it as well as anyone equipped with only the human senses could. And then he lived through the hell of seeing that bond destroyed.

In Ohio, there is growing resistance. More people are realizing that local communities don't benefit from injection wells, for example, and that the only beneficiaries other than the drilling companies are the well owners and the state, which receive fees paid on the injected waste. Recently, several Ohio communities, including Athens, Broadview Heights, Oberlin, Yellow Springs, and Mansfield, have enacted laws to control drilling. Said the Athens city council president: "We're very much concerned about protecting our water supply."

Those Ohioans concerned about fracking in their communities, however, are learning that they have little control. Fracking decisions are made by a state legislature full of people whose campaigns were heavily financed by the industry, and these decisions are enforced by a state Department of Natural Resources that often seems more concerned with protecting industry than the natural resources under its watch. At a recent ODNR open house on fracking, for example, a few dozen mostly middle-aged academics from nearby Kent State were met by fourteen armed ODNR police and a police dog. One man in his eighties told a reporter he had never been to a public meeting "so oversupplied with armed people."

In February 2015, the Ohio Supreme Court, in a 4–3 vote, ruled that cities and counties can neither ban nor regulate fracking through zoning laws or other restrictions. Wrote one dissenting justice, "What the drilling industry has bought and paid for in campaign contributions, they shall receive."

Perhaps most significantly, fracking decisions in both the legislature and courts are generally made by people who don't actually live on the ground where the fracking takes place. Literally separated by geography from having to see or hear or smell the effects upon their native habitat, they can more easily separate themselves from the consequences of their decision. It is, after all, a "rational" decision, usually justified as bringing more jobs, more tax revenue.

Yet as the saying goes, there's no free lunch. The cost of those jobs, that revenue, will be paid by someone (the poor) and something (the earth), somewhere (anywhere but here).

We in the West live in a civilization shaped by the European culture of Enlightenment. It was, and arguably still is, a culture based on dualism, on fragmentation, which requires that in order to gain knowledge we must separate subject from object, male from female, body from spirit, human from nature, and so on. Our culture is based on the assumption that, as Fred Kirschenmann writes, "we could achieve certainty only through such separation. This doctrine of separation taught us to see the world in fragments instead of relationships. And that changed everything!"

I have been thinking about separation ever since I began this book, our separation from the ground by pavement, our separation from the grounds that give us our food, water, energy, and spirit. If we are to avoid the worst kinds of hell that seem to threaten our future, perhaps it's this separation we must address.

In an essay titled "You Are Accepted," published in 1955, the Christian theologian Paul Tillich argued that because "sin" had become such a well-known word, it had become "strange" to us—it had lost its power. His goal was to recapture that power, and the word he chose was "separation," as in "sin is separation." But while Tillich does use the word "separation," he goes further—in a sense, to sharpen our understanding of what it means to be "separate from"—by using the word "estrangement."

Does the distinction between the two words matter? According to my old friend David Saetre, a longtime professor of religion at Wisconsin's Northland College, it does. "The distinction matters because estrangement is, I would argue, a more radical term," he tells me. "So we're estranged from God, we're estranged from one another, and we're estranged from ourselves. Sin is not the naughty things that we do; it's the condition in which we experience ourselves."

Remarkably, Tillich describes this estrangement as an estrangement from "the Ground of Being," a phrase Kent Gramm had used while talking with me about Gettysburg. In fact, in his writings Tillich often referred to God by using this term. God is not a being, Tillich argued, but instead the ground from which all being derives. It's the idea that God is beyond definition, that anything we say about God is not God. As soon as we start claiming qualities of personalities to a deity, we're already missing the point.

What especially fascinates me, as I consider what I have seen so far, is how Tillich's use of this phrase might relate to the literal ground. "If we find ourselves in a condition of our existence where we are estranged from our source," David explains, "then literally we become estranged from all other facets of being, including then the ground upon which we stand. We pave it, we cover it, we weave rugs for it—we do everything we can to not stand on ground. And Tillich would say, 'Well, yes, of course.'"

My friend tells me this reminds him of the Adam and Eve myth, how everything they do in that myth is an acknowledgment of this estrangement. They hide from God, they cover their bodies—there is a physical barrier placed between the self and the other, the self and the deity, and the self and nature. As Tillich wrote, "Because we are estranged from the Ground of our being…and we do not know where we have come from, or where we are going. We are separated from the mystery, the depth, and the greatness of our existence."

In his sermons, David tells me, he will sometimes use the words "We come from the ground, and we will return to the ground," knowing that his audience will understand him to mean "ashes to ashes." But, he says, that's not the same thing. "We come from the ground" is consistent with the Hebrew meaning, he explains. The name "Adam" is a play on words; Adam (*adama*) is Hebrew for "earth." Adam and Eve are made from the ground. In other words, in the Christian faith, we all are made from the ground.

And here is the separation that Tillich talks about, the hell Pope John Paul II referred to as "the state of those who freely and definitively separate themselves from God, the source of all life and joy." We come from the ground but now are separated from the ground. We are distanced from the source of our being. And once we consider ourselves separate— like an Ohio lawmaker voting to frack another part of the state, or a Minneapolis business owner overusing pesticides on the corporate lawn he never walks—our actions no longer feel as though they have a direct impact on us. "The problem with this otherworldly emphasis is that not only are we estranged from the earth from which we came, but it takes the next step and says that it's not our destiny," my friend explains. "We don't really belong here. In its worst form, then, ecologically, we can do any damn thing we want because all of this is fallen. Our destiny is not the earth. Our destiny is instead a heavenly home that is earthless, soilless, groundless."

Yet this is not what I have learned from the ground, from the soil, from the beauty of the earth I have witnessed in the sandhills, in Bishopstone, on an evening in Gettysburg when fireflies rose from the battlefield where my fellow Minnesotans fell. Everything I have learned so far about the ground tells me that we are not separate from the rest of creation but in powerful ways connected. This is, after all, the lesson from the newest sciences—including the science of ecology, which has as its founding tenet that everything is connected to everything else. Or, as John Muir said of the Yosemite, "When we try to pick out anything by itself, we find it hitched to everything else in the Universe."

Returned from Ohio, I have lunch with a friend and tell him about my trip. "What more can you say about fracking that hasn't already been said?" he asks. And in many ways that's true. Anyone with an Internet connection can find plenty of statistics and articles claiming both bonanza and doom. But what you can find only by going to where fracking is happening is what it feels like on the ground.

"There are no unsacred places," writes Wendell Berry, "only sacred and desecrated places." I admit that what I experienced in Ohio brought the word "desecration" to mind, but only in the loose sense in which we use the term—severely damaged, wiped out, ruined. In fact, of course, the actual definition refers to treating "a sacred place or thing with violent disrespect." The vital meaning of the word—the reason it is more powerful than the others—is this specific reference to the sacred.

Soon after I leave the state, the Ohio legislature—while working on a bill to speed up the permitting process for fracking—agrees to hold off opening the state parks to gas and oil exploration. It turns out that state forests are already open to fracking, as are state wildlife areas. And the sense I get is that in time, even the parks will be fair game, that with its friends in government and industry, and enough voters either tuned out or benefiting from the current situation, this occupation will spread whenever and wherever it desires.

I think about the parks and forests and wildlife areas near my home, and what I would do if this occupying force I saw in Ohio threatened my lakes and loons and rivers. I would like to think that I would chain myself to tanker truck tires, or pour sugar into pickup gas tanks, that I would monkey-wrench pump jacks and burn down man camps and in every possible way join the resistance.

But resistance against whom? Where does this occupation start, when does it end? Faced with the desecration of the grounds on which we base our lives, what would any of us do?

Treblinka

Scientific models... cannot tell us all we need to know.
— David George Haskell, *The Forest Unseen*

In the middle of eastern Poland, in a clearing among pines, the train track ends. The numbers tell what happened next: between July of 1942 and August of 1943, a mere thirteen months, the Nazis murdered more than nine hundred thousand people here. And the "here" is important: the Nazis chose this place because it was in the middle of nowhere, and they wanted somewhere secret, a place where every person sent there was sent there to die.

But murdering these people—mothers, grandfathers, children clutching favorite stuffed bears—was not enough. The Nazis then tried to erase their lives by burning their bodies, turning human stories to ash. And even that wasn't enough, because when the Nazis realized that the war would be lost, they attempted to extinguish the story of the camp itself—and they almost got away with it. With the Soviet army steadily advancing, the Nazis razed the buildings, tore up the train tracks, tilled the ground, planted wildflowers and pines, built a farmhouse, and left.

I arrive on a beautiful day in April, thinking that surely I have reached hell on earth. Though at least in the case of Treblinka, hell looks remarkably like a county park, an opening in the forest with closely cropped grass surrounded by the living greenery of pines. I have worn cleats and chased Frisbees on grounds that look like this: bunches of tiny yellow flowers, clumps of greener grass, fallen tan leaves. Outside the boundaries of the former death camp is a small museum, and here in the clearing a single Soviet-era concrete memorial statue surrounded by jagged stones symbolizing the many villages, towns, and cities from which the murdered were taken. But unlike the far more often visited Auschwitz, Treblinka has no barracks remaining, no gas chamber ruins. There is very little to help you imagine what it was like here, which makes it, in many ways, even more terrifying.

Remarkably, for more than sixty years everything we knew about Treblinka came from the testimony of a few survivors, Nazi guards, and a single article by a Soviet journalist who visited the site after the Nazis departed. Then, in 2007, a team led by the British forensic archeologist Caroline Sturdy Colls began the first-ever archeological work on the site. She wanted to find further proof of the Nazi crimes committed here, to help silence those who today deny that Treblinka was anything but the "transit station" the Nazis told their victims it was. She knew that despite the decades that had passed, and despite the Nazis' best efforts to erase the evidence, "the landscape could never be sanitized in that way." In large part to respect the actual ground, the team used several new technology-based observation techniques, including LIDAR, a method in which lasers measure the distance between the earth's surface and a plane overhead, thus essentially stripping away the vegetation to reveal the contours of the naked ground. Sturdy Colls and her colleagues looked for elevations and depressions that might mark the locations of previously unknown mass graves and gas chambers.

Having identified several potential sites in the forest opening, the team made four small excavations. The first two produced some interesting relics, including a fossilized shark's tooth that revealed that the Nazis had dumped sand from a nearby quarry here to help disguise the ground. But it was in the next two excavations that the team found what they had suspected might be there: a brick wall and foundation. They knew from archival documents and the limited accounts of survivors and witnesses that the only brick buildings in the camp had been the gas chambers. They also found broken orange tiles of the type the Nazis had used on the floor of the gas chamber to further the lie that it was just a Jewish bathhouse, and artifacts such as hairpins, rings, and brooches, some pressed into the earth near what would have been the gas chamber doors. "It's horrible to think that all you leave behind on earth, to be found seventy years later, is a pendant or a hair clip or an earring," said Sturdy Colls. "It makes you think if the whole earth of Treblinka was opened up what kind of hell you'd be looking into."

Before leaving for Poland, I contacted the Dutch archeologist Ivar Schute to find out more about the new archeological techniques being used at Treblinka and elsewhere. Not only does Schute work with Sturdy Colls on her Treblinka project, he also works at Sobibór, another little-known Nazi death camp in eastern Poland that claimed among its 250,000 victims some thirty-five thousand Dutch Jews. In fact, the three extermination camps in eastern Poland—at Treblinka, Sobibór, and Bełżec—were part of the Nazis' Operation Reinhard, a systematic murdering operation that was incomprehensibly lethal; of the nearly 1.7 million Jews shipped to these camps, barely more than one hundred survived the war.

The overwhelming lack of firsthand testimony, combined with the Nazis' attempt to erase the evidence of their crimes at the three camps, makes understanding exactly what happened in these hells extraordinarily difficult, he told me. But the new remote-sensing technologies such as LIDAR, along with developments in geophysics, geographic information systems, and digital archeology, are allowing archeologists and historians a much improved understanding of where the murdered took their last steps. In Treblinka, for example, the team found what they suspect are the gas chamber foundations some eighty meters from where historians had thought they must be, and at Sobibór located what they are certain are gas chamber foundations underneath a Soviet-era asphalt parking lot.

The new archeological technologies are also particularly important at these camps because, Schute explained, in addition to the mass graves, "those sites were slaughterhouses, and there were random killings everywhere." With the mass graves, the burning of bodies, and the random murders, attempting to piece together burial locations is incredibly difficult. Ground-penetrating radar is able to detect changes in the properties of underground materials, producing scans in which, for example, soil looks different than bricks or sand. Another technology, electrical resistivity tomography, already commonly used in oil and gas exploration, gives archeologists yet another way to see into the ground. (At a site in Lithuania they recently used it to discover a long-rumored escape tunnel dug by Jewish prisoners.)

Still, Schute said, while the new technology can help archeologists understand the ground in ways not possible even ten or fifteen years ago, "in the end you still have to dig in some places to really know what it is." In Sobibór, right beside the gas chambers, he told me, "we found an area with gold teeth and dentures. I think you understand what happened there. A place like that is not a structure, it's not a building, it just happens inside that camp, and things like that you can only pinpoint in an archeological way. It's a little weird to map gold teeth, but that's what you do as an archeologist."

In Treblinka, Schute explained, "The camp still exists, it's all there. As an archeologist, the camp is still a place you can visit. There is so much in that soil." When I asked how far down he and Sturdy Colls had had to dig before they began to find evidence, he says, "Something like fifteen centimeters—it's directly under the surface." The Nazis tried to demolish the camp in order to disguise their acts, he said, "but they never thought about archeologists, because in the ground it's still there."

At Treblinka, the goal is much less to dig than it is to locate what remains in the ground and to leave it in the ground. Schute described the digging so far at the camp as "test plots of two square meters, not more than that. More like peepholes, you might say." As much as they found the first time, he told me that there is "lots more to discover yet," but he expects they will again primarily use noninvasive techniques to pursue their goal. "Look," he said, "it's kind of a sacred ground."

I asked what he meant. "That's difficult to answer. Because of my work I come across human remains, and I deal with lots of families of the victims. I also spoke with the survivors, so it's quite difficult to be unmoved by that. My whole perception of those places is highly influenced by those contexts. You have to be careful, because digging is destroying, and after you've dug it out, it's not there anymore."

*   *   *

Another person whom I spoke with before leaving the States was Shiri Sandler, director of the Auschwitz Jewish Center at the Museum of Jewish Heritage in New York. A cheerful, serious woman in her early thirties, Sandler comes from a family of Holocaust survivors and has been to eastern Poland dozens of times—though, she told me, "Treblinka I've only been to once, thank God." We sat in the museum's conference room, its windows overlooking New York Harbor, the Statue of Liberty in the distance.

Auschwitz has become the symbol of the Holocaust, she told me, because it was so enormous and so many people died there, but also because it was a place that people could survive. "You couldn't survive Treblinka, and so we don't know what it was like," she explained. "It's impenetrable. It's lost. I had family in Treblinka, and the idea that this is a space that erased these people completely, in a way that Auschwitz didn't, that's really horrifying."

I asked, Can we ever imagine what it must really have been like? "I think we all do it," she replied, "but I don't think we can ever actually do it. We could in theory understand what it feels like to put your foot down one after the other between the ramp at Auschwitz, or the ramp at Treblinka, and the gas chamber. We can understand what it feels like to walk that distance. But we would never understand what it feels like to be starving while doing it, to have the terror while doing it, to have been separated from your child, your parents, to be beaten, to smell the smoke. We cannot access that. And yet of course we try. My old boss used to say you're not a scholar at Auschwitz, you're not a doctor, not a lawyer— you're a person, you're a parent, you're a child—it breaks down everything else. And so we're always going to be putting ourselves in their shoes, but we're never going to do it correctly."

I wondered too, to deal with the horror, whether we sometimes imagine that those who suffered the Holocaust weren't really like us. "Every once in a while it hits me, the reality of what it must have felt like for my grandmother to lose every single person she had loved in the world," she said, "and that's just simply too much. We can't imagine that kind of loss. Our brains can't comprehend it, so we think it must have felt different. They must have loved their children less, or their hunger didn't feel like what we think hunger feels like. We think that somehow it didn't feel for them like it would feel for us, when in fact it was exactly what it would feel like for us."

On the day of my Treblinka visit, my guide Josef has already taken me to the sites in the Polish capital where the last Warsaw Ghetto fighters were entombed, the Germans burying them in the rubble. And the hospital site where the Germans killed all the staff and then blew up the building, burying five hundred patients in concrete. We have visited the deportation site where the Germans shoved Jewish families onto cattle cars and sent them east. And we have driven for what seems like hours. As we get closer to Treblinka, the road begins to twist and disintegrate. Even in our comfortable van we must slow, and turning in to the forest we slow further still. As we draw near, the forest becomes individual trees, and I wonder whether any of those I see were here back then.

We park at the small museum and see Jewish gravestones used by the Germans to pave dirt roads, and a diorama of the camp showing naked bodies burning on the railroad-tie grill. A handful of photos line the walls: the steam shovels used to dig mass graves and later to dig up the bodies to be burned, the Polish Jewish doctor—Janusz Korczak—who cared for orphaned children in Warsaw and chose to die with them here. There is a color print drawing of a young girl holding a songbird upside down, dead in her hand. And there is a young girl's voice, maybe the daughter of the Polish woman who took our entrance fee, echoing through the rooms. As we stand by the diorama, Josef tells me that Treblinka is where he stopped believing in God. "If he is Almighty and he allowed this to happen, he has a twisted sense of humor and I prefer not to believe in him."

And then we begin walking.

Walking is the way to approach the site, your feet on the ground. We pass the opening in the trees marking the boundary of the camp, six-foot-high stones placed fifty meters apart showing where the barbed-wire fence would have been. We pass the stone marker telling the camp's story in Polish and English, saying the memorial was created between 1959 and 1964, and I think, *So this site stood empty for fifteen years after the war, just a clearing in the forest where the devil had lived for a year.*

I will never forget the quiet. Josef says I am lucky there are not busloads of Israeli tourists, that I am lucky to be alone. And I agree—this is what I wanted. But I think too of Shiri's words, that we are always alone when we visit these sites. No one can experience this for us. Next we walk without talking to where train tracks lay. The wind in the pines, our boots on the gravel path. And all around are sounds I do not hear—no constant rush from a distant highway, no jets passing over, no radios or televisions, no sounds from what the world has become. Back then, though, it must have been so loud.

"Of course," says Josef. "There were thousands of people, and the dogs of the guards, and the shooting to the air or shooting to the people. The train was going in and out. The crying of the kids and the screaming of the women." He pauses, "And the smell probably was horrible."

When we reach the train platform, the path turns to cobblestone, the stones with their multitudes of shapes, their rusts and grays, greens and blacks. It looks like the memory of milling feet—here had stood hundreds of thousands of people, having just stepped off the train, once again under the sky, between rows of pines, trying to make sense of where they were. They were on earth, still, for a little more time. And maybe here is where they were closest to what they had been, for the last time. Here, these embedded stones mark the place where these people still wore clothes from home, the men still in ties, the children in their best shoes, the women brushing hair from their eyes.

The long stone platform merges into a walkway, and this is where the people left the disembarkation platform and entered the camp. The large Soviet-made mushroom-shaped monument looms over the center of the site, and the jagged shards of stone rise from paved ground. I stand thinking that here is where so many others once stood, having just arrived, scared, not knowing what will happen next. Josef says he will wait for me here, that I can walk around the site, and so I enter alone.

Treblinka was a murder factory, a dis-assembly line designed with ruthless efficiency and thoroughness. Jews arrived in train cars, mainly cattle cars, having been told they were being resettled to Ukraine farms, many having spent days packed so tightly they had to stand. To further the deception, the Nazis built the train platform to look normal, complete with a fake ticket window and signs pointing directions to various cities. A small orchestra sometimes played. Upwards of twenty thousand people were "processed" through each day.

We barely know what happened at Treblinka. Nine hundred thousand people were murdered on this small spot of ground, and we will never know all their names, never have their bodies to bury. Our best record of what took place comes from the Soviet journalist Vasily Grossman and his twelve-thousand-word article that begins, "Let us walk through the circles of the Hell of Treblinka."

From the train platform the people were herded through the camp gates and onto a large square, three to four thousand at a time. The women and children were ordered into the barracks to undress. "Love—maternal, conjugal, or filial love—told people that they were seeing one another for the last time," Grossman writes. But they were hurried along by the SS. The women had their hair cut. The men were told to undress quickly and stack their clothes neatly. A terrible stench was in the air, and the smell of lime chloride. One of the few survivors of Treblinka remembered a young woman asking how long it would take. Ten minutes, she was told. Looking around she said—as though to the world, to this life—"Farewell."

Now the naked people were ordered to give up their valuables, their identity cards, money, wedding rings, earrings, jewelry. Now the guards tore wedding rings from fingers, ripped earrings from lobes. Anything deemed valuable for the German war effort was saved, the rest of the documents tossed away as trash. Back in the empty square their belongings were being rifled through, everything of value to be sent to Germany, everything else buried or burned.

Now the naked people were ordered into a column five persons across and marched down an alley 120 meters long and two meters wide, barbed wire on either side, and lined with SS men. This is the path the Nazis called *die Himmelstrasse*, "the Road to Heaven." Grossman reports that this path was "sprinkled with white sand, and those who were walking in front with their hands in the air could see on this loose sand the fresh imprint of bare feet: the small footprints of women, the tiny footprints of children.... This faint trace in the sand was all that remained of the thousands of people who had not long ago passed this way, who had walked down this path just as the present contingent of four thousand people was now walking down it."

The walk took about a minute. When they saw the gas chamber doors, the people stopped in astonishment, until the Nazis drove them through using rifle butts, biting dogs, lead pipes. There are stories of resistance—a young man stabbing an SS officer, a young woman grabbing a gun and killing two of the SS. Stories of mothers protecting their infants, their children. Stories of peasants in the nearest village escaping into the forest to get away from the women's screams.

Most of the victims were dead within twenty minutes, with the children lasting longest. Any survivors were shot. A team of workers armed with dental pliers then pulled the gold from the victims' mouths, and the naked bodies were dumped into a mass grave.

In the winter of 1942–43, the Nazis began to burn the bodies of the murdered. They also dug up the previously buried and burned these too. After struggling with the logistics of how to burn so many human bodies, they settled on digging an enormous pit some 250 meters long, 20 meters wide, 6 meters deep. Over this pit they placed a grill made of railroad ties. At night, in surrounding villages, the flames could be seen above the pines. In the daytime, the ashes of the bodies were taken out of the camp and dumped in the woods. Local peasants were forced to load their carts and scatter the ashes on the road leading away from the camp. Children were made to spread the ashes more evenly with spades; car wheels made "a peculiar swishing sound on this road."

In the autumn of 1943, having done their best to hide the hell they had created, the Nazis left. Grossman arrived about a year later, describing a quiet scene with the pine trees along the railroad tracks "barely stirring. It is these pines, this sand, this old tree stump that millions of human eyes saw as their freight wagons came slowly up to the platform."

And then he describes the earth's surface . . .

. . . casting up fragments of bone, teeth, sheets of paper, clothes, things of all kinds . . . half-rotted shirts of those who were murdered, their trousers and shoes, their cigarette cases that have turned green, penknives, shaving brushes, candlesticks, a child's shoes with red pompoms, embroidered towels from the Ukraine, lace underwear, scissors, thimbles, corsets, and bandages. Photographs of children from Warsaw and Vienna, letters penciled in a childish scrawl, hundreds of perfume bottles of all shapes and sizes — green, pink, blue . . . and thousands of little forest flies are crawling about over all these half-rotted bits and pieces. . . . We walk on over the swaying, bottomless earth of Treblinka and suddenly come to a stop. Thick wavy hair, gleaming like burnished copper, the delicate lovely hair of a young woman, trampled into the ground.

Seventy years later, I stand looking down, wondering whether this is that spot. I think of the bodies burning day and night. I think of the carts filled with ash. I try to figure where the so-called Road to Heaven was, knowing that somewhere nearby is ground that supported the nine hundred thousand murdered as they walked their last steps. Though I have been assured by Shiri and others that walking here is okay, it feels almost disrespectful. Because the bodies are nowhere, they are everywhere.

When you go to Auschwitz and Birkenau, of course it is terrible, but in a different way, because of the barracks and photographs of victims and gas chamber ruins. Here, there is almost nothing. Even the jagged stone markers seem to have risen from the earth of their own accord. At Auschwitz, I connected with two photographs, one of a Polish writer staring defiantly at the camera who was my age when he was murdered. And the other of a thirteen-year-old girl with tears in her eyes, just months before she died. I imagined lying on the barracks bunks on a frozen winter night, and I imagined walking down the ramp into the gas chambers, which are left as they were after SS dynamite. I imagined stepping off the cattle car that remains at the selection point. Of course you can't know what it was like, but the buildings and ruins and photographs help your imagination along.

At Treblinka, there is little help. In the middle of the forest clearing lies a long rectangle of black stones in a slight depression in the ground, a monument to the pyre where the Germans burned the bodies day after night. The grass around the memorial is worn to dirt from visitors' feet. But there are no words inscribed, no photos, just black rocks that disturbingly resemble charred remains. And here is where my imagination fails. Now, dear brain, I would like you to imagine hundreds or even thousands of naked human bodies roasting—it's said women burned more easily than men because of their larger fat stores, the stomach of pregnant women would finally burst, spilling the fetus—imagine this, please, imagine the hiss and splat of bodies split, of fat falling into flames, the stink and stench of burning flesh, the blackest smoke. I'm asking you to imagine hell as you stand at the memorial's edge, and you are saying, No, I'm sorry, I can't. Or perhaps it is No, I refuse.

What's it like to walk at Treblinka? I only know what it was like for me, early on a spring afternoon, a warm sunny day. And I know it is where, more than any place I had been, I had a sense of sacred ground. I say this not feeling certain that it has much to do with "sacred places" as defined for us by others—a cathedral, for example. And that maybe anywhere we call sacred ground has as much to do with mystery as it does with being sure.

"Though we may often be told to 'look up' to find God ... the truth of creation is that God is near ... in the soil beneath our feet," writes Norman Wirzba, professor of Theology, Ecology, and Agrarian Studies at Duke Divinity School. When I return from Poland, I look him up and ask him about sacred ground.

When we use religious language, he tells me, "that doesn't necessarily mean that we know what we mean. Because to enter into the ground is precisely to enter into the unknown. But we end up going in the direction of religious language because the reductive, scientific terms that we have just aren't up to the task." It's a sentiment I have heard from the best of scientists, an appreciation that while science with its methods and models can help us comprehend the world and what's happening to it, science cannot, as the biologist David George Haskell writes in his beautiful book *The Forest Unseen*, "tell us all we need to know." All the digging in the world, all the latest technological tools, will only tell us so much about the ground of Treblinka. There is something there that cannot be scanned or measured, and yet no doubt exists.

How then to make sense of the experience of being there?

"In the context of that kind of incomprehension," Wirzba tells me, "the only really responsible reaction is a kind of humility that is steeped in acknowledgment, and the role of a human being as a witness to something that exceeds our power of understanding." It's the role, he says, that asks of us attention, respect, gratitude, and celebration. And it is the opposite of a human being who is, in his words, "divorced from" and "bored with" the true sources of sustenance. It is the same way, he explains, that rather than appreciate the mystery and priceless importance of the natural world, we are given "virtual worlds in which you couldn't be bored, because the explosions keep getting more spectacular."

This reminds me of my choice of the word "ground," as opposed to "earth" or "land." It's as though we have become bored, I tell him, with the ground at our feet. What is there to be amazed by, or wonder at? "Ground" is a word that is almost like a disguise, because it's so plain. And yet...

"Everything depends on the ground," he jumps in. "It seems to me that in our culture, we're at a loss to know how to determine the significance of anything." In order to help his students to understand, as he says, "the soil as the ground of signification," Wirzba gives them the semester-long assignment to grow a plant. Many of them haven't got a clue how to do this. "They've never touched soil, and it's amazing to see the transformation that happens when they really have that *aha* moment when they recognize how soil is this complex living reality," he says. "We don't know if soil is an organism or a matrix or what-the-heck-is-this-thing, and yet when we put seed into it, it can support life."

For students, Wirzba tells me, this experience then becomes "the basis for being able to talk about the value and significance of things that is real in a way that just reading about them isn't." And that, having had this kind of experience, "people can come to the sense that a place is sacred, or ground is sacred, because they begin to see how the power of life that makes their life possible is at work here."

This is what happened for me at Treblinka, and it's a feeling that has grown since that time. I have come to believe that whatever is sacred is that which inspires in us an understanding of the connections that matter. I mean the connections that sustain us. I mean the connections—the web of relationships—that keep us alive. The power of Treblinka is that in being there you have the opportunity to be stripped bare and helped to realize your connections with other human beings. You know that on this ground people were literally stripped bare, stripped of everything, until the one thing that was left were the bonds between lover and loved. Husband, wife, father, mother, child. This is something no Nazi hell could break, and perhaps the "it" that no technology will ever discover.

Alaska

Never before this had I been embosomed in scenery so hopelessly beyond description.

— JOHN MUIR, *TRAVELS IN ALASKA* (1915)

Alaska: the word means "wild," and so much of it still is. For one thing, the state is almost incomprehensibly big. A three-hundred-thousand-square-mile fire, one that would envelop any city in the lower forty-eight and burn it to ash? Here that same-sized fire can be burning and almost no one will know. Like few places on Earth there is wildlife here, grizzlies and wolves, a caribou migration some hundred thousand strong, and vast open ground where few people have been and fewer go. More than four thousand miles from my last stop in Poland, traveling from that old dark history to this long summer light, I have come to walk the wildest ground I can find.

Where I'm going, in the southwestern section of the state, in the vast Yukon-Kuskokwim Delta Wildlife Refuge, birds fill the skies—more than a million ducks and half a million geese breed here every year. More than one hundred thousand swans, forty thousand loons, and thirty thousand cranes return to the refuge to nest each spring. As for shorebirds such as black brant, emperor geese, cackling geese, and Pacific greater white-fronted geese, the refuge is the most important nesting area in the country. The refuge alone, at nearly twenty million acres, is larger than several US states, larger even than Maine, and the second-largest refuge in the National Wildlife Refuge System. The Bering Sea wraps around its northern, western, and southern shores, and Alaska's largest rivers — the Yukon and the Kuskokwim — flow through its body. Seals, whales, walruses; bears, caribou, and moose — the waters, grounds, and skies are alive here, every season around the year. There are other wild grounds I could have gone to in the world, but what's here is still incredibly wild, and here I have a friend.

We met in New Mexico fifteen years ago but haven't seen each other for ten. I left for Nevada, and Shane came here, following a girlfriend who was following a job. He had grown up in Iowa and looks and acts the part, blond and blue-eyed, happy-go-lucky, kind. His girlfriend lasted less than six months here and left Shane behind, but by that time he'd found a good job himself, working with Yup'ik kids in the school, and he decided to stay. Eventually, he met and married Carolyn, herself half Yup'ik, half white, and together they have three kids. She works in the schools and he works at the public radio station. And he hunts, something he'd never done before he came here. But he has found out who he wants to be, who he is, you can tell. He's married into this life and made this life his own. He's the same Shane I knew in New Mexico, but now he's a hunter and a father, and he has learned this ground.

"It starts with desire to explore," he says. "And like in New Mexico, if you wanted to explore you could get in a car. But here it's not quite so simple. So you have to start to learn the native terrain. There's no road to take you there."

You need to know that in order to survive?

"Yeah, unless you want to stay in your house, or just go to the grocery store." But the groceries here are expensive, priced to reflect the reality of flying containers across hundreds of miles of roadless land. So the Yup'ik culture survives by hunting and fishing and gathering berries. The culture survives as it has for thousands of years through what we call subsistence, though there's no such word in the Yup'ik language. In Yup'ik, hunting and fishing and gathering berries means staying alive.

"So you go with people who know," he continues. "You go out and watch. That's a big thing in Yup'ik culture. Don't ask, don't talk. The first time you're seeing something, just watch."

As Shane talks he lays out some common foods: strips of smoked sockeye salmon, a jar of moose meat, a bowl of berries mixed with Crisco and sugar called *akutaq* but known as Eskimo ice cream. There are also beach greens eaten with seal oil, the foulest-smelling food I have ever encountered. Shane and Carolyn have a walk-in pantry well stocked with canned salmon and moose, and freezers full of caribou, goose, ducks, whitefish, halibut, salmon, and berries, all the result of Shane's work. He's learned a way of life that has long been expected of Yup'ik men, of *yuuyaraq*, which means "the way of being human."

This way of being human has survived for more than twelve thousand years. The most numerous of the three indigenous groups in western Alaska more commonly known as Eskimo, the Yup'ik are descendants of the hunting-gathering peoples who came from Siberia. Numerous, though, means something different in this part of the world; the entire population barely reaches thirty-five thousand. Villages of fewer than a thousand are scattered throughout the sparsely populated (at least by humans) territory. Even in Bethel, the main village, the population adds up to only six thousand. No roads lead to Bethel, or anywhere else in this part of the world. The only practical way in is by air, and the only way out into the hunting areas is by water.

In the morning we load the fishing boat with plenty of extras—food, water, clothing—and guns. When Shane goes back into the house to get one last load, I stare at the rifles and shotguns set leaning against a seat. We are going out in a boat with guns. It's something I never do, but I have asked for this, I have asked Shane to take me hunting. He knows I haven't shot a gun since I was a boy in northern Minnesota, and he knows that I doubt I can shoot at any birds now. But he's patient, not judging. He's been where I am and knows. What birds are out there, wild and alive, that will end the day dead in our boat? Shane comes back outside just as an elderly Yup'ik man weaves his way down the muddy rutted road on a child's bicycle. "Good luck to you guys," he says.

We pick up Shane's friend Charles, a baby-faced, full-blood Yup'ik in his twenties who grew up with his grandparents in a small town a forty-five-minute boat ride from Bethel. He didn't learn English until he was a teen, and even now he's mostly quiet, though when he does speak he does so with a smile. "Which one of you is going to use my twenty-gauge?" he says as he hops in the cab and looks back at me.

We drive to the boat landing and then motor out for what seems like hours. We stop for lunch—smoked salmon, canned moose ("That's like the backbone and stuff," Shane says when I point to the strangeness in the jar), and Charles's favorite: spreadable ham on Pringles. I ask Charles what it's like to hunt for so much of his food.

"I was pretty squeamish at the start. The first thing I killed was a muskrat, and it looked so cute. That very first time was the hardest. Used a twenty-two when I was nine. But I guess after times like that I got used to this being my way of helping out." We are sitting in the fishing boat, which Shane has driven into the river's muddy shore to hold us fast. The land is long grass and knee-high plants, no concrete for hundreds of miles around. Birds on the horizon seem to know to stay away.

"It's how I grew up," Charles continues, "so nothing about it seems strange. What does seem strange is when we got a teacher that had never been hunting before." That would be me. Or Shane, when he first arrived. The white guy from the lower forty-eight who has never had to kill to eat, at least not directly.

"What did your grandfather teach you about respect when you hunt?" Shane asks his friend.

"I guess leave as small a footprint as possible," Charles says quietly. "The first thing I was taught is that there's a time of year to get certain things, spring, summer, fall. And you could say in a way it's disrespectful not to take advantage of those— there's a time of year to plant, and a time of year to fish, and a time of year to hunt birds. I guess it's taking advantage of those whenever they become available. I guess if this is what you have to do."

This—shooting animals—is something I have never had to do. Could I do it? Of course I could, if I had to. But I understand that culture and necessity make hunting a way of life here, and that in our living we kill all the time—we just don't see that killing firsthand.

"It felt natural to me," Shane says when I ask, "and it felt difficult. I remember the first caribou I took down, the first animal I killed here. I remember it being a powerful experience. I really can't draw up those emotions now, it seems foreign to me that I would feel that sad about it. I remember a really heavy feeling in my body. I shot it from a hundred and fifty yards, and I had a scope, so it was a clean kill. It didn't suffer much. I remember thinking, *It's so big, like, what a big life to end.* I remember its size really affected me."

When you got close to it? "Yeah. I remember I was like, 'I can't believe I killed this much life.' And then that kind of came and went. So I really like to get clean kills, I mean a lot more than hitting an animal and letting it suffer. Because, if you really start to think about that animal as a being, and if you believe it has its own consciousness, then it's pretty hard to think about inflicting that kind of pain. I still feel it with birds, especially when I'll take one and not the other."

You've done that before? "Yeah. With swans, they'll stick around, they'll hang out. Like one, it'll fly to the other end of the lake and just wait. They die hard too because they're strong, so sometimes you bring them down and...I've gone on long chases where I've been glad I'm in good shape. Because they can run on the tundra about as fast as we can. And so I'm closing on it slowly, over a quarter mile, half mile, and then catch it."

And then what? "Kill it, shoot it. Shoot it in the head. Hopefully as quickly as possible. Soon as I can get close enough. That doesn't feel good either. Prolonging its life until that. But yeah, its mate will just sit there and watch the whole thing. And you know it's just praying to God—or, at least that's how I imagine—just wanting its partner to come back. I hate that. So now I don't shoot at them unless they're close, unless I can get both of them, because I just can't stand to take one and not the other."

*   *   *

We continue on for another half hour by boat, finally landing and loading backpacks. Shane and Charles have extra-large packs and hope to fill them with dead birds. But while I'm happy to be with my friends, I'm not going to share in their shooting. To be witness will be enough, at least today. I can already tell that just being on this ground—the tundra—will be like nothing I have experienced before.

For one thing there still are no roads, nor even any paths. We tromp west toward a ridge rising maybe two hundred feet, on the other side of which Shane and Charles know lies a shallow pond and hunting blinds. They have had good luck there in the past, and so we slog along in our rubber waders and suspenders, sweating in the waterproof material, carrying heavy loads. And we do so on soft ground covered with vibrantly colored leaves—purples, reds, greens—and berries of blue and red and yellow. We each regularly kneel to grab a handful of blueberries, our hands stained by their juice. Cranberries, salmonberries, blackberries. In some places I find it impossible to avoid stepping on these small lovely leaves and berries and feel as awkward as I would stomping through a neighbor's flower garden back home. Moss, lichen, shrubs—below these rusty, heather hues is a ground made of matted plants and branches and roots, and you feel like you're walking on ground made of sponges, as though elevated above the "real" ground—the hard ground you're used to—as though walking on top of a forest with branches tightly wound, a canopy that must cover an undiscovered world. And in fact, beneath this soft colorful tundra lies a frozen ground—permafrost—a vital element of this ecosystem that is in ever-increasing danger.

*   *   *

Permafrost all around the Arctic is in danger of thawing as a result of a warming climate. In Alaska, permafrost underlies some 80 percent of the state, and already the state's citizens are dealing with sinking roads, shifting foundations, and the disappearance of wetlands and lakes as the ground beneath them thaws and the water drains. Forest fires burn hotter and longer on ground drier now as a result of less permafrost. "Drunken trees" lean and topple as the ground in which they have grown loses its frozen grip. The Yukon-Kuskokwim Delta is at the southern boundary of permafrost, and is a place where scientists are working overtime to understand the severity and consequences of the thawing.

It is hard to overstate the importance of their subject. As a recent report from the Woods Hole Research Center stated, "Carbon emissions from thawing arctic permafrost will become substantial within decades, likely exceeding current emissions from fossil fuel combustion in the United States." And because these emissions from thawing permafrost build on themselves in what's known as a feedback loop, "the potential exists for a catastrophic, self-reinforcing cycle of warming and thawing permafrost" that could lead to "out-of-control global warming."

Almost unbelievably, current projections for "acceptable" levels of carbon emissions—including those used to frame the Paris Agreement in 2015—do not, say the Woods Hole scientists, "adequately account for carbon loss from thawing permafrost, so current projections of future climate tend to be too optimistic." That is, the frightening scenarios we have heard about climate change caused by fossil-fuel emissions don't even include one of the biggest contributors.

And how much carbon are we talking about? A recent

model projects that 130 to 160 petagrams (a measure of weight) of carbon might be released from thawing permafrost between now and the end of the century. To put that in perspective, the current annual carbon emissions for the entire United Sates is about 1.4 petagrams. In short, adding the carbon emissions from thawing Arctic permafrost is like adding the emissions of an additional industrialized nation the size of the United States to the world's annual total carbon budget.

What's hard to understand is how something that is so potentially catastrophic could gain so little attention from the public, or at least from our elected officials and policy makers. I speak with Dr. Susan Natali, one of the Woods Hole Research Center scientists who works in the YK Delta (as it's known) and contributed to the new report. She agrees that the looming threat of thawing permafrost is "not good news, but it's worse to not know about it, to not pay attention to it. Having knowledge is progress. You can't do your bookkeeping if you don't know that you have a leak."

Natali points to President Obama's visit to Alaska in August of 2015 as a hopeful sign that perhaps people are at least beginning to pay attention. "If we do nothing," the president said, "temperatures in Alaska are projected to rise between 6 and 12 degrees by the end of the century, triggering more melting, more fires, more thawing of the permafrost, a negative feedback loop, a cycle—warming leading to more warming—that we do not want to be a part of."

Natali describes the process of permafrost thawing as like "taking a chicken out of the freezer." That is, it takes a while. But once permafrost thaws, microbes decompose it, and this decomposition process converts carbon-rich organic matter into greenhouse gases, carbon dioxide and methane. The frightening thing, she explains, is that "once permafrost thaws there's nothing we can do to stop decomposition because it's a natural biological process. Microbes break down organic matter in the soil to gain energy." She pauses, then adds, "It's so huge, the Arctic, and there's so much permafrost. Even though the rate of thawing and decomposition may seem relatively slow, you multiply that by this large area and it becomes critically important."

Critically important, indeed. The potential impact of thawing permafrost is almost impossible to imagine. Twice as much carbon is stored in the frozen Arctic tundra as has been released since the industrial revolution of the mid-1800s. These are soils that have been frozen for thousands of years, and now we have come along to release them. They are a Pandora's box to beat all others—who knows what will happen? Of course, no one really knows, but Natali and other researchers point to the range of permafrost thaw by 2100 as being between 30 and 70 percent. "That range is not because there's uncertainty in the physics," she tells me, "that's because there's uncertainty in human behavior. A thirty percent thaw is if we completely cut fossil fuel emissions; seventy percent is business as usual."

Even with such uncertainty, what probably can be figured already is that thawing permafrost will mean fewer wetlands, fewer plants, fewer animals and birds, fewer lives. In fact, the Yup'ik are already seeing this happen in their wild world— fewer berries to pick in the summer, fewer animals to hunt in the fall and winter, less ice and snow—green grass in January, bees in February, pussy willow blooming three months early. As one fifty-four-year-old mother here recently put it, "There's so much not normal stuff going on."

*　　*　　*

We reach the blind to find two Yup'ik hunters arrived ahead of us, both about thirty yards down the shore, hidden in tangled branches. Spent shells in artificial greens and blues and yellows litter the ground. The two Yup'ik men share friendly smiles, their faces smeared with bug juice. One leans out of the blind to prop the heads of dead geese on sticks for decoys, then whispers a loud "Shut the fuck up" to his partner as more geese appear in the sky nearby. I crouch behind Shane and Charles, hoping the birds will turn away, but they don't. Six curve in flight over the pond, and four hunters rise like mechanical pistons to blast away. Remarkably, only one goose is hit, the other five curve away crying, higher and higher, their V broken. Trying to stay with the group, the hit one flaps hard but falls to the water with a splash and starts a frantic one-winged swim toward the pond's far side. One of the Yup'ik begins trudging around the shore toward where the goose disappeared. They can be difficult to find and are referred to—these wounded birds, hiding from the hunters in the long grass—as "fox food." If the hunter doesn't find them, they will certainly be found later tonight.

I leave the blind, walking a couple of hundred yards back to where the ridge drops away. Behind me spreads unending tundra, below me a wide valley. As I did at One World Trade Center and as I have at other grounds, I take off my shoes. To the soles of my feet the uneven, spongelike surface is soft, soggy, and cool. The ground is colored with the golds and reds and blues of berries, the colors vivid as though rising from deep wells. I sink to above my ankles, my feet disappeared. And yet my footing is firm, anchored. Five brown cranes appear in slow big-winged flight. I urge them away with berry-stained hands. On the horizon hangs virga, the rain that doesn't reach earth. And then, silence.

When I was writing *The End of Night*, some of my best experiences came when I found myself somewhere with no artificial lights in sight. Most of us are so used to the presence of such lights we don't notice them anymore. To look around and realize there are no lights anywhere — nothing on the horizon, reflected off the clouds — can be breathtaking. It's an experience of finally escaping the modern age. You fall back in time to when this was life — no electric lights here below, a bright star-spread above. I have that same feeling here on the tundra, with no evidence of the human anywhere in my sight, no human sounds reaching my ears, just endless unbroken ground to where the gray sky and green horizon meet.

I lie back on the soft tundra and close my eyes. In an early scene from McCarthy's post-apocalyptic novel *The Road*, "the man" remembers how "Once in those early years he'd wakened in a barren wood and lay listening to flocks of migratory birds overhead in that bitter dark. Their half muted crankings miles above where they circled the earth as senselessly as insects trooping the rim of a bowl. He wished them godspeed till they were gone. He never heard them again." This short passage has for me always been one of the most heartbreaking in that heartbreaking book. We humans have left the birds nowhere to land. Here on the tundra, everywhere is someplace to land.

I turn over and lie stomach down, arms spread, making an indentation in the blanket of small grasses, plants, leaves, and berries, then I sit up. A single crow circles above, the effortless shuffle of wings in air, he croaks and flaps off, maybe to report my presence. Strange to sit where maybe no one has. I remember now how it is out here, you have to be still and wait. Like eyes adjusting to dark. Time is different. Nothing seems to change, barely even to move.

I am thinking of my first night in Alaska a few days ago, fishing on a river that did not seem to move. I could tell we were drifting only by watching the banks glide past. We had strung the net across the silty brown river, one end anchored, the other end held by Shane in the boat's bow, waiting for the fish to hit the net, entangle their gills twisting and fighting, lose their lives as Shane hauled them toward us and flipped them into the white tub in the bottom of the boat. I manned the engine, keeping us turned toward wherever Shane said, and stared at the tub. Dark dried blood splattered on white plastic walls, evidence of earlier fights, earlier deaths, earlier lives ended.

The river is more mirror than see-through—if I fell in I would instantly disappear, the quiet world going on above. It's hard to believe that schools of salmon fill this river—a hundred yards wide here—that great groups of fish move toward us, toward the sea. There might as well be whales or submarines going by below us, houses and buildings unhinged going past, the dirty surface concealing whatever worlds go by below. I wonder—

"Hell, yeah!" Shane shouts. "Here we go."

First one, then another, then another again. As the fish hit the net hung across their path the white bobbers hop and sink, the fish fighting against the net, flashing silver sides. Shane hauls the net toward the boat, camouflage rubber waders and a red baseball cap, straining against the weight. He pulls the first salmon from the water and flips the two-foot-long body into the tub. The fish flops slippery against the white plastic, sliding one wall to another while spraying blood from its mouth and gills. Shane flips another fish into the tub and the two flop on top of each other, eyes wide as though shocked, uncomprehending, their mouths shaped as if frowning. They soon tire and lie gasping, now and again their bodies flipping, one final try, then lie still. Shane wasn't sure we would catch anything, or perhaps only a few, but now he's afraid we will catch too many and that some may go to waste. The salmon lie smeared with each other's blood as though in marinade, the stench growing as the tub fills. Shane shrugs and tells me that after you've been in beaver guts, nothing compares. "You wouldn't think it would smell so awful," he says with a laugh, "it's just trees!" He's working as quickly as he can, sweating, just a T-shirt under his suspenders, dragging the dead weight of drowning fish, the net like a giant necklace of lace and thrashing jewels. I am a mix of emotions, mostly stunned. I want to pray, to give thanks, to ask forgiveness. Somehow this doesn't seem right, to pull life from water and watch it die, and yet these fish will feed us—and the village, as later we drive around Bethel handing out salmon to anyone Shane knows, following the rule that "You've got to find a home for all these fish," knocking on neighbors' doors to offer "as many as you want." And tonight, Shane will sever heads and slice open bodies to remove the sacks of bright pink pearls, the eggs—and carve long pink filets that taste wonderful grilled.

But for right now, they lie gasping in the tub, small fins still moving, dying. Shane takes one, says it's chummed out, won't taste good anyway, and holds it delicately in the river, one hand under its belly, the other taking its tail and drawing it back and forth, moving water through gills until the body begins to sway and he lets it back into its old life.

With fishing, with picking berries, with hunting, I know that this is how it is here — the natural world giving life to the Yup'ik, directly, as it has for thousands of years. I think again of Norman Wirzba's words, that we come to sense a place — or ground — as sacred when we "begin to see how the power of life that makes [our] life possible is at work here." I know I'm up close to it now, to the reality of our reliance upon other life for our own, rather than removed, as I too often am. On the wild ground of southwestern Alaska, I realize I have never been anywhere before that so allows me to see this connection.

I rise and wander back to the present, to the theater of shooting, just in time for five geese to come flying over, curving their bellies at the four hunters rising from blinds. This time all five go down, splashing, two not moving, two others still flapping, one making for the far shore grass. Shane immediately sets out, walking in shin-deep water, taking maybe a hundred yards to catch it. When he reaches the goose he grabs it by the neck and twirls the body, then drags it back to the blind. The two Yup'ik claim three of the geese, the two other birds lie at our feet. The yellow legs and toenails — the guidebook calls them greater white-fronteds, the Yup'ik call them yellow leggers — yellow webs between each yellow toe, the tiny white teeth, the necks narrow, stained with blood.

We wait a while longer, then leave without any more kills. As we hike back across the tundra, Charles walks ahead as I ask Shane what he's seen in hunters here, whether they respect the relationship between human hunter and other life. "It's an individual thing," he says. "I've been out with hunters that have been like textbook adherence to the values that we've been taught. And then I've been out with other hunters who are just like fucking Rambo. It's like, We're fucking killers, and that's what we came here to do. Like one buddy of mine, every time we catch a moose, he was taught by his father to cut a tip of its heart off and bury it to the east. But at the same time if he's got his automatic rifle and sees a caribou herd, he's not going for headshots. He's just like, 'I'm going to make sure I kill something,' and so he's just like *pow-pow-pow-pow-pow-pow-pow-pow-pow*, he's putting five bullets in an animal to make sure it comes down. Whereas I would just never do that."

He continues, "I've seen it both ways. When we're ice fishing we'll string a net under the ice right when it freezes up. All the whitefish and burbots, they start running. And this one time, we were catching some, but this other Yup'ik guy was just downriver, and he was picking the hell out of his net. And so we went over to talk to him—asked him, What are you doing to get so many fish? And me and my buddy, we knew all the basics about how to respect your catch, give it away. But for like a half hour he just talked to us, all about how you treat your animals, and the land around you, and the people—give it away, show respect, don't waste any, make sure elders get it—it was just this litany. And my friend, he can be a smartass sometimes, said, Yes, yes, aside from all the metaphysical things we can do, can you give us some advice on, like, where we should put our net?" Shane laughs.

"So that was a guy who took it seriously. But then you know, you'll come with someone else and they're like 'Let's fucking shoot seagulls.' Both Yup'ik guys, two different mentalities."

It's then that the flock of sandhill cranes turns in the distance and begins coming toward us. If they keep their direction, they will fly right overhead. Shane and Charles immediately crouch. "Get low," Shane whispers, sets his pack on the ground and cradles his rifle. I can tell that both my friends are excited—there are no other hunters around, so these birds will be all theirs. I, on the other hand, am thinking how this is pretty much the last thing I want to see, sandhills shot from the sky, dead at my feet. Earlier in the day I had told Shane that I can't believe anyone shoots cranes, that I could never do it. He laughed and assured me they taste wonderful, the "rib-eye in the sky," some say. Now I stare at the ground and listen as the trilling crane calls drift our way.

As Charles and Shane ready their guns, I remember a story Shane told me the day before, of how he and another friend had been out hunting in the winter and got separated on the way home. "We were hunting on the snow," he'd told me, "we were hunting ptarmigan on this creek six miles from here. And on the ride back my buddy, a white guy, came over a hill, and right there, twenty-five yards away, it was like swan-swan-swan-swan-swan-swan-swan-swan-swan sitting there, and he didn't kill them. And if he had told that story to our Yup'ik friends, they would have been like, 'You disrespectful asshole. They presented themselves to you, and you just did nothing. They were giving themselves to your family.' But he said, 'Dude, that low yellow sun had just cracked through the clouds, and it was glowing. And I wasn't going to pull my gun out.' But we had just spent three days killing animals."

I'm thinking of that story when, just as the cranes—at least twenty—are about to come into range, they change course and take their slow-winged evening flight away from us. And as my two friends relax their grip on their guns, I hear Shane whisper, "Beautiful."

The Sierra Nevada

Marriage is a coming together for better or for worse, hope-
fully enduring, and intimate to the degree of being sacred.
— William O. Douglas (1965)

T he ancient animals are still available—jaguars, turtles,
the night flight of bats. The mountains—the Sierra
Nevada de Santa Marta—are still full of life, one of the world's
most biologically diverse ranges, eighteen hundred species of
flowers, seventy species of bats, fifty species of frogs, and liz-
ards, tortoises, caimans. More than six hundred species of birds
migrate through or live here, many found nowhere else on
Earth. "And if there is an earthly paradise in these Indian
lands, this seems to be it," wrote the Spanish friar Pedro Simón
in 1627. "It is completely crowned by high peaks....From all its
ridges flow streams of golden waters that tumble like crystal
snakes from the high peaks deep down into the valley."

The two Arhuaco Indians, Wilfrido and Atty, indigenous
Colombians dressed in white cotton robes and—on Wilfrido—
the white cap honoring their snowcapped-peak home, have
come down from this heaven to meet with me in the Museo del
Oro Tairona–Casa de la Aduana (the Museum of Gold) on the
grounds of their ancestors in the seaside city of Santa Marta.
Founded by the Spanish in 1525, the city was built atop the
ancient home of the Tairona, the people who lived here from

roughly AD 900 to 1600, and the Nahuange, who lived here in the seven centuries before that. As we view a display of pottery found beneath the museum during its remodeling, Wilfrido says, "Maybe if they dug more they would have found more of our things."

As in countless other places the history here is one of "discovery" and destruction fueled by a lust for gold and other spoils. The arrival of the Spanish meant the beginning of the end for the Tairona, who fought the invaders until the turn of the century. When in 1600 the Spanish finally defeated the Indians, they sentenced sixty indigenous leaders to death, burned entire villages, and chopped the body of the Indian chieftain Cuchacique into pieces and placed the mess on display.

And yet, from blood and ashes arose the culture represented before me. Wilfrido, forty-something, and Atty, barely into her twenties, speak their own language to each other and Spanish to me and my hosts. The Arhuaco are one of four tribes that descended from the survivors of the 1600 destruction who fled into the mountains above Santa Marta. Along with the Kogi, Wiwa, and Kankuamo, the Arhuaco share a determination to express to the rest of the world their connection with nature.

"They know me as Wilfrido, but that is a borrowed name," he says. "I was born at dawn, and that is my name." His long black hair falls tangled from beneath his white knit hat. He nods at Atty. "Her name is the name of a sacred site, a lake at the upper part of the mountain." She stands barely five feet tall, with long black hair and canvas tennis shoes beneath her long white smock. Both carry hand-woven black-and-white bags. "The girls represent Mother Earth, and we represent trees," Wilfrido continues. "Without them we cannot exist."

For the Arhuaco, the Sierra Nevada range is peppered

with more than 360,000 points that when connected form a "black line" that signifies the heart of the world. Understanding these points is the job of their "spiritual fathers," the Mamus. Wilfrido is studying to become a Mamu, a rigorous years-long initiation into a deeply intimate relationship with nature. For the Arhuaco, writes the ethnologist Wade Davis, "every element of nature is imbued with higher significance, so that even the most modest of creatures can be seen as a teacher, and every feature of the world mirrors the whole."

"For us all things are sacred," Wilfrido explains. "Everything that our eyes see, that we can visualize, is sacred. Natural things. Of everything that is sacred, there are some points that are more sacred, that are of paramount importance. It is in those sacred points that the natural laws are written. The indigenous person in the Sierra Nevada, especially the Mamus, we know those ancestral laws, from the rules of the trees, of the animals, the water, the snow, the sea. A Mamu learns from the Mother Earth, from the trees, from the river, from the mountains. That is the university where the Mamu learns."

There are no holidays in their society, no religious calendar. Their experience of time exists only in relation to what is happening in nature. "Our festivities are when the rain comes, when there are a lot of animals, when the plants start flowering, and the flowers will come, and the bees will come and the small animals will start sucking the nectar from the flowers. When there is equilibrium between the snow and the sea, when there is harmony between what is down here and what is up there. When there is harmony between the cold and the heat. When the life and death understand each other. When day and night are harmonized."

Foremost among the Arhuacos' belief is that their care for and attention to nature in the Sierra will guarantee the health of nature elsewhere—that their spiritual work and their way of life are responsible for maintaining Earth's ecological balance. "Every morning, every time you see the sun, every time you see the moon, every time you see the stars, you have to feel grateful to the gods," Wilfrido explains. "Every time you have the opportunity of drinking water of a natural source, not from a pipe, of looking at the snow, of sharing with an animal, you have to thank those gods and you have to feel grateful for that."

"We have," he offers with a smile, "a very deep connection with the earth."

Listening to Wilfrido, I'm reminded of Joanna Eede writing in *National Geographic* about her experience of sitting with a member of a previously "uncontacted tribe" in Brazil. "Our grandparents lived here," he told her. "I am part of the land. Without the land, we have no life. This place is my home." Eede was humbled "to imagine how strong the sense of belonging to a place a people must feel after 10, 20, or even 55,000 years rooted to one part of the Earth." She quotes Davi Kopenawa, a spokesman of the Yanomami people in Brazil, saying, "It is hard to describe how connected my people are to nature. You can't uproot us and put us in another land; we don't exist away from the forest." It's a thought, Eede wrote, echoed in a Cherokee statement: "We cannot separate our place on the Earth from our lives on the Earth, nor from our vision and our meaning as a people."

I don't mean to suggest that any of us in the modern industrialized world can instantly achieve a "very deep connection with the earth." But we ought not to think it impossible for a human being to have such a connection.

*   *   *

"I want to talk about the gold," Wilfrido says as we face a display case featuring dozens of small gold figures—jaguars, turtles, bats. "These pieces here were a sacred representation. We do not think about the commercial representation of gold. It was just a way of representing a figure, while at the same time paying spiritual tax—as you pay the light, the Internet, the phone—to the gods of those animals. For that reason, when the Spanish came, we were giving away the gold. It doesn't have a commercial value to us."

To say gold has a commercial value in our world is obvious understatement. It is rare, yes, and beautiful, but also just a mineral mined from the ground. Yet this mineral has driven us mad. Maybe the most powerful discussion of this madness for gold—for material wealth—is *The Rediscovery of North America*, a slim book by the American author Barry Lopez. Lopez takes us back to the moment Christopher Columbus "discovered" America, describing how Columbus stole from the sailor who first saw land the promised prize for doing so, and how "what followed for decades upon this discovery were the acts of criminals—murder, rape, theft, vandalism, child molestation, acts of cruelty, torture, and humiliation."

It's hard, I think, to comprehend the viciousness of the Spanish toward the indigenous peoples they found—the dismemberment, beheading, and rape, of how they "cut off the legs of children who ran from them. They poured people full of boiling soap. They made bets as to who, with one sweep of his sword, could cut a person in half. They used nursing infants for dog food." We have seen this kind of unimaginable cruelty in our own time from the Nazis, most notably, but also from the Khmer Rouge in Cambodia, from ISIS, from an ever-growing list. The tragic power of Lopez's book comes from his connecting of the Spaniards' vicious acts spawned by "a ruthless, angry search for wealth" with an approach to the world that is alive and well today. Nothing has changed, he argues, in our attitude toward the natural world or toward those people whom we see as "other" than us. We still approach the natural world with, as he puts it, "the assumption that one is due wealth," and no amount of riches is ever enough. We hunger for more, even as the Earth shows in countless ways that this approach cannot continue.

Yet perhaps even more powerful is his positive nod toward the future, his offering a vision for how we might go forward.

> This violent corruption needn't define us. Looking back on the Spanish incursion, we can take the measure of the horror and assert that we will not be bound by it. We can say, yes, this happened, and we are ashamed. We repudiate the greed. We recognize and condemn the evil. And we see how the harm has been perpetuated. But, five hundred years later, we intend to mean something else in the world.

For Lopez, this "something else" begins with "looking upon the land not as its possessor but as a companion...to cultivate intimacy, as one would with a human being." We would, he writes, "have to memorize and remember the land, walk it, eat from its soils and from the animals that ate its plants. We would have to know its winds, inhale its airs, observe the sequence of its flowers in the spring and the range of its birds."

We would, in other words, have to learn to approach the world in a way similar to that of the Arhuaco, similar to that of indigenous peoples around the world, similar even to that of those Western traditions and thinkers that have sought to direct us toward a different relationship with the earth than the one we've been trained to accept.

Kathleen Dean Moore, author and professor emeritus of philosophy at Oregon State University, is well versed in these traditions, Western and beyond. In books such as *Riverwalking*, *Wild Comfort*, and *The Pine Island Paradox*, she uses personal stories and experiences as starting points for her reflections on our relationship with the natural world. She is especially interested in action, in what we can and should do, and why.

In "What It Means to Love a Place," from *The Pine Island Paradox*, Moore takes a rowboat into the early morning fog of the Alaskan bay her family visits every summer to consider the question her title implies. She actually makes two lists, one for what it means to love a place, the other for what it means to love a person. And to her surprise, the lists are the same. The list goes from one to ten, and the first nine come easily, beginning with "To want to be near it, physically," and "To want to know everything about it." Her list includes fear of loss, determination to provide protection, and desire: "To want the best for it." And then she stops, feeling there is something missing but not knowing what.

Then she writes, "Loving isn't just a state of being, it's a way of acting in the world. Love isn't a sort of bliss, it's a kind of work." Number ten, she writes in her notebook, is "To love a person or a place is to take responsibility for its well-being."

A week later, I stand in another Sierra Nevada, this time amid the Mariposa Grove in California's Yosemite National Park. Here some five hundred mature giant sequoia trees, the largest organisms in the world, face a decidedly uncertain future, threatened by our warming of the planet. The deep brown trunks rise from pine-needled floor like the resting hooves of some enormous animals. Some bases are cracked, some hollow, some almost black, fading upwards into a red-brown trunk. They are like trees made for some larger planet, set here during creation and forgotten. And yet they are ours, the limbs on the largest like trees themselves, growing horizontally, curving up or down or both, like an eight-armed acrobat in some circus show. From a distance, the people at their feet are like ants with cell phones and hoodies, some with cigarettes, the smoke curling into mist.

It's hard, maybe impossible, to fathom the trees' age. John Muir reported that "the age of one that was felled in the Calaveras Grove, for the sake of having its stump for a dancing-floor, was about 1300 years, and its diameter, measured across the stump, 24 feet inside the bark." He reasoned that "unless destroyed by man, they live on indefinitely until burned, smashed by lightning, or cast down by storms," and that he "never saw a Big Tree that had died a natural death; barring accidents they seem to be immortal." Another time he spent all day figuring the age of one that had been cut down, crawling on hands and knees to count "with the aid of a pocket-lens." He eventually counted four thousand rings, "which showed that this tree was in its prime, swaying in the Sierra winds, when Christ walked the earth."

I am joined in Mariposa Grove by Gus Smith, a Park Service fire ecologist. Gus gained Internet fame in 2013, when he was featured in two short films about that summer's Rim Fire, California's worst forest fire ever, a fire that burned more than 250,000 acres, including 100,000 within Yosemite National Park. I'd heard that in some places the fire burned so hot it essentially "killed" the ground, and indeed of the

acres burned, more than 40 percent burned at "high inten-
sity," meaning that not only was all the vegetation killed but
all the microbiota in the top inches of soil were killed as well.
This kind of destruction, says Gus, is something we almost
certainly will see more of in the future.

"All the predictions are that we will have bigger fires," he
says. "The fire season is getting longer, fires are getting hotter, and they're burning more intensely." So intensely, it
turns out, that they are stumping the computer models firefighters use to fight them. "These fires," says one California
official, "are actually exceeding what our models will even
predict."

When the fires come—hotter and fiercer and more
frequently—Gus doesn't have the option of evacuating. "My
job is to drive back up the hill and deal with it," he says.

"Have you ever been in a situation where you were afraid?"

"Oh, yeah," he nods. "And I watched a tanker pilot die
this summer. It was just brutal. I was a mile away, and I saw
the belly of this plane, and I saw the sunlight hit his wing,
and then I saw the other wing kind of parallel with it—in a
way it shouldn't be, and I was thinking, what am I seeing?
And then a big black cloud of smoke and fire. I was in tears."

And death, of course, so much death among creatures
that cannot move fast enough to escape—all the animals,
reptiles, insects for whom the ground is home. And death
among the firefighters too, as in the 2013 Yarnell Hill Fire
that left nineteen "hotshot" firefighters killed.

"It's usually heat in the lungs," Gus explains. "When you
inhale the heat, it's basically like a burn. And burns weep. So
the insides of your lungs weep and you suffocate, you drown."

We walk in silence, on a day of rain and few visitors, the
ground soft, soggy, the giant trees standing as they have for

centuries. We stop by the Grizzly Giant, the hugest-fattest-thickest tree you can imagine — or, rather, even huger-fatter-thicker than you can imagine. I often think about how nothing I ever see in movies or science fiction or television matches the beautiful design of what we already have in nature, from the small — like the stained-glass wings of a dragonfly — to these enormous trees. They are so enormous, in fact, that I'm finding I cannot fit them into my camera's frame. Every photo I have taken is of a tree's base or, looking skyward, the crown. The Grizzly Giant, the Bachelor and Three Graces — even when I back away a hundred feet I still can't hold them.

There is only one place where these enormous beautiful trees grow, and that is right here in this Sierra Nevada, several dozen small groves scattered along a narrow 870-mile band on the west side of the range between 5,000 and 8,000 feet. That's it. Nowhere else on Earth. They have evolved to grow in this particular ground.

It doesn't seem possible that these enormous trees could disappear. There are little ones everywhere around the grove, gatherings of small sequoias, spindly fluorescent bluish-green, just starting out in life — and the giants have been here for so very long. But the drought currently strangling California may keep the small trees from developing the robust root system they need. Worse yet, ecologists say that if the drought continues, in a century most of the giant trees could be gone. It turns out that these and other giant trees, some of the oldest living beings in the world, are among the first organisms to feel the pressure of climate change. "We are talking about the loss of the biggest living organisms on the planet, organisms that play a key role in regulating and enriching our world," says Australian scientist Bill Laurance. "A world where a child can't stare up in wonder at a giant cathedral-like crown is a very real possibility."

Unfortunately, the California sequoias aren't the only big trees in danger. Recent studies indicate that around the world the biggest, oldest trees are in danger from higher temperatures and drier conditions. The journal *Science* found "rising death rates among trees 100 to 300 years old across a wide range of global landscapes." Craig Allen, a top USGS tree expert, confirms that "globally no major forest type is immune to episodes of drought- and heat-induced mortality," and that it's happening "from the Amazon to Alberta and everywhere in between."

How strange it is to walk among these giant trees and try to imagine them gone. Then to think that it's our choice to burn fossil fuels like the fracked oil and gas I saw being produced in Appalachia that might have a hand in the erasing. To love these trees that take your breath away is easy. But to then, as Kathleen Moore writes, take responsibility for their well-being? At some quick point the threat feels overwhelming, the task beyond our control. To turn away feels like an understandable reaction. And yet, here we are in this still-beautiful world that with its every sensory gift asks only that we respond.

Miles Silman, a Wake Forest University biologist who has worked in the Peruvian Amazon for more than a decade and seen the unrelenting change brought to that wildest of grounds by climate and population growth, knows this dilemma well. A man with an ever-present smile, Miles is passionate about his science, but he's equally passionate about his role as father to three grade-school-age children. He knows that the world they inherit will be dramatically different from the one we know now. He recently returned from

South Africa, where he experienced a new kind of zoo, a national park enclosed by fences, with the animals contained and where there's no longer natural migration. Instead, "every lion's branded—it's been caught—and then they ship them from park to park," he tells me. "So they don't have migration corridors anymore, they have FedEx."

My question for Miles is, What keeps him going when it would be easy to lose hope about the future of a world he obviously loves? "One of the points that I have been making in my talks recently—and a lot of what I have to say to people is a brutal downer, there isn't a whole lot of good news—is to remind people, after you've shown them all the things we are doing to the planet and all the things we will have to do to feed eleven billion people and keep from driving everything else extinct, is that things get better on the other side. You think about all the things we're doing—increased temperature, disassembly of ecosystems, deforestation—all this is taking the world as we know it and passing it through a bottleneck. But the world doesn't end in 2100. Eventually, the population will come down, temperature will come down, and all of the pressures that human consumption puts on the planet will go away. We'll decarbonize the economy. Then the question is, What world are you left with?"

There have always been challenges in the world, Miles says, but the difference with the changes we are causing now is that, unlike events such as wars, in which the effects are mainly felt by humans, this time "We're taking down the rest of the world with us. We're taking out all the megafauna, and we're going to take out, depending on how narrow the bottleneck is, up to half of the species on the planet. That's different, because that doesn't come back. We'll come back, but the rest of it relies on what we take with us."

And that, Miles says, is where the actions we take now matter. Anything we can do to keep that bottleneck wider will help shape the world we get on the other side. While right now humans are exceeding nature's capacity, that won't always be true. Eventually, fertility rates will begin to decline. By 2200 or 2300, for example, estimates are that the human population may well be back to seven billion people—or even far fewer than that.

Of course, just as we can't know exactly how much carbon the Arctic's thawing permafrost will release, we can't know the exact number of our future population. But we know the direction we're heading. According to the most recent UN projections, from 2015, "there is an 80% probability that the population of the world will be between 8.4 and 8.6 billion in 2030, between 9.4 and 10 billion in 2050, and between 10 and 12.5 billion in 2100." In terms of population, the future numbers all depend on people's behavior, Miles tells me. For example, once developing countries acquire Western levels of wealth, will they then adopt the very low Western birthrates? Already, at least, the world's population growth rate is falling, and some studies predict the world's population will fall as low as one billion in the year 2300.

Whatever the exact number, assuming that the human population does indeed fall below today's level, "All of a sudden, things don't look so bad anymore, right? They're not so dire," he says. "The question then is, What have we taken with us when we get back to that point? What can we bring through? On some level, you're setting up the world as the ark, and the flood is what we've unleashed on ourselves. I like the thought of looking at the world as an ark," Miles continues, "because at least it gives you agency."

Here is where our purpose lies. There's every reason to do

everything we can right now to preserve and conserve the world's natural wealth and beauty, because we know serious change is coming, right? "Yeah, it's on us. It's what we're going through, but it's still at the wide part. The question is, How narrow is it going to get before it broadens back out again?"

I am reminded of talking with Susan Natali about thawing permafrost, and her saying that at least it's better to know. Denying the looming threats won't mean they will disappear. But if Miles is right, at least we can get to work doing all we can to soften the blow.

How to do this? My sense is that there are countless ways. I am more convinced than ever that our literal and figurative separation from the ground beneath us—and the unlimited potential of sacred ground for connecting us to what matters—is key. I think of Norman Wirzba telling me, "Theoretically, at least, any place, any ground, is sacred because all ground, in some way or other, contributes to the life that's happening above it, making it possible. There's no reason that my raspberry patch isn't also sacred ground. And to realize that that raspberry patch is sacred in some small way means that I now have to relate to it in different sorts of ways."

But what if we have no raspberry patch, or any immediate access to anyplace—let alone a Treblinka or an Alaska—where we might feel the unmistakable sense of the sacred? What if, as Wirzba described, "our suburban life, urban life, automobile life, pavement life, all of that, puts the barricade between us and that sort of experience, and we can't experience the places of our lives as sacred?"

What if we have nowhere we came from, or nowhere we want to return?

Home

From this one place I would explore the entire north and all life, including my own. I could look to the stars and feel that here was a focal point of great celestial triangles, a point as important as any one on the planet. For me it would be a listening-post from which I might even hear the music of the spheres.

— SIGURD F. OLSON, *LISTENING POINT* (1958)

The best flights come in from the west, sometimes circling around the city beforehand, and pass over ground I know well. My parents' house, the elementary school I attended, the golf course where Luna and I would walk on winter nights accompanied by owls, deer, and coyote. I always try to be seated next to a window on the plane's left side, so that I can find these and the others: stadiums for the Twins and the Gophers, streets lined with maple, ash, pine, and elm, and the lakes in a chain heading south—Cedar, Isles, Calhoun, and Harriet. Sometimes, at dusk in autumn, a moon will rise behind the city with its beams atop the river, the Mississippi a silent wild current curving through a million or more lives. And then Lake Nokomis, just blocks from the house, and the parkway leading to my steps, where if this year is as it was last a finch will soon live in the eave above the front door. I am forever grateful for the opportunity to travel, forever in awe of airplanes in flight (I've had it explained to me plenty, and still…), but I am always happy to be returned to these grounds where I grew up, where my family lives, and where the rest of my life is now taking root.

Back in Minneapolis to stay for a while, I decide to visit a man who, as an artist and art historian, has spent much of his life engaged with the sacred. Opening the door to his basement studio, Wayne Roosa assures me that "whatever I think of the sacred is just cheek and jowl next to the quotidian." I see it immediately in his current project, a series of small wood shapes that evoke tombstones or monuments and yet are made with hacksaws and scrap pegboard. The sacred and the everyday together.

When I ask him first about landscape painting, thinking that this might be a particular form in which artists have explored the relationships between humans and nature and the sacred, he nods and mentions John Constable, the English Romantic painter from the eighteenth and nineteenth centuries. "His landscapes, you could almost walk by them, they're just countryside scenes. But if you stop you realize he's given you little paths that lead you deeper and deeper in." And then, Roosa explains, you notice there are all these patterns of light and shadow on the ground. "That's because he's painted the sky, the clouds, in movement. And what you're really seeing down here is the light breaking through from the shadow of the clouds. So even though the ground is a stable element that's apparently fixed, and the sky is the transient element that's in motion, he's playing those two off each other, and he's using light and shadow to make you realize that this is somewhat ephemeral." On one hand, Roosa tells me, Constable's landscapes can seem like little more than a stroll through the woods. "But at the same time, the way he paints reminds you that while the ground may seem like a static place, ultimately you're temporary here."

Thinking about our lives in relation to nature and how this might push us toward an understanding of the sacred has long made sense to Roosa. A trim man in his sixties with salt-and-pepper hair, in blue jeans and a brown T-shirt, he grew up in Wyoming, a place of "big land and small houses," where his deepest impressions were of "land that goes on and on, and then this huge sky." He describes a childhood and youth in which he had "a deep immediate relationship with nature" that became the background from which he then moved into culture and art. Of his generation he says, "We were deeply imprinted by nature. And then we moved into culture and human construction of artifacts and artworks. But the grounding was in nature, and even our sensibility of what metaphor is and how it works harken back to this existential grounding in nature in our childhoods." As his career progressed, he began working with younger artists in New York City who told him they couldn't relate to his experience. Instead, he says, "They talked in terms of simulacra, that their relationship with 'what you call nature' is mediated through culture and kitsch and cities and urban experience." Roosa leans back in his chair and grins. "I would say, 'Let's go to the Boundary Waters together.' And they'd say, 'Yes, I would like to visit nature.' And they weren't being irreverent. It's just that nature was not a primary reference or grounding for them. It was something you went outside New York City and visited."

Still, Roosa talks about his younger peers with admiration. "The whole local movement has arisen from that," he says. "People looking for something that feels authentic, when the whole world seems like a fabrication of layers of politics and power plays. They can't affect that, but they can plug in locally. They're growing a garden on a rooftop. And their diet is really focused on what they eat and where it came from." Roosa tells me there's a movement among contemporary artists they're calling "social practice." That's what they're after, he says, trying to infuse the local, the daily, the ordinary to make us realize that it has an aesthetic component. And while some of them would see that as sacred, others would be far more materialistic. "But the point being," he says, "to make us value what we're doing immediately, now, here. Because otherwise, we're ruining the world."

A few years ago, Roosa helped coordinate a show at the Minneapolis Institute of Arts simply titled *Sacred*. Ten different curators were each given a small gallery and told to put together their own exhibits. Some curators chose costumes or medieval manuscripts, Roosa says, while another put together a show arguing that "the little knickknacks that we all have on our shelves, like your kid's third grade baseball trophy, are sacred." For Roosa, the gallery that connected most was the one on walking. In this gallery, the curator had featured everything from the classical—Chinese landscape paintings where "you'd see the landscape in all its mystery, and there's always a little guy with a walking stick, and he's the old man on a spiritual pilgrimage walking through nature"—to the work of a contemporary artist using an entirely modern medium. Conceptual and performance artist Stan Shellabarger took a twelve-by-twenty-five-foot roll of red rosin construction paper, stretched it over diamond-plate metal, put on a roofer's work shoes, and walked in a labyrinth pattern until the little diamond plates wore through. "So on one hand it was nothing but cheap construction material that had been walked on by a worker, the mundane," says Roosa. "But at the same time it was a labyrinth, and he was gearing it off prayer meditation paths. He was trying to say these go together. And part of his premise is that our whole life, we are doing this. Unconsciously we're walking a labyrinth whether we're meditating or just trying to make money."

Then, says Roosa, there's Francis Alys, another artist who in several performance pieces has opened a can of paint, tipped it upside down, and let it drip as he walked — through the streets of Paris, Jerusalem, São Paulo, and elsewhere. "He's saying it matters where you walk and why you're walking, so I'm leaving a trace," explains Roosa. "Something as mundane as where you walk in life, if you pay attention to it, you realize there's meaning there. It makes us aware of how we're here, and how we behave while we're here."

"Meaning" and "sacred" are not synonymous, Roosa adds, "but they are inseparable."

And so, I'm thinking of walking, of this book's first pages in Manhattan and all the walks since, of how walking brings together the sacred and the mundane. And I am thinking of my friend Derek, who may never walk again.

Before I moved to Reno for graduate school, we were housemates in Albuquerque. To my dog Luna, he was "Uncle Derek," and Luna's big sister was Derek's Bailey, a wiry golden retriever. For three years the four of us shared the house at 2428 Pueblo Bonito: Pudge Brothers pizzas with pepperoni and green chile, the fire pit out back with Scorpio in the southwestern twilit sky, Derek's candy-colored climbing ropes above the back room's Saltillo tiles, the sliding glass door with two dirty dogs waiting to come back inside.

Derek left New Mexico first, moving to Seattle, where he met Amanda, his perfect match. They were constantly camping, climbing, and biking, constantly outside. I have never known anyone as positive as Derek, and Amanda's the same. Stepping into their house, I had to raise my level of gratitude. Then the terrible news arrived on Facebook, the most efficient way for Amanda to reach Derek's many friends. "This is a very difficult message to write," she began. On a snowy February Friday, skiing in northern New Mexico, Derek had crashed in a way that made it almost certain he would never again feel the ground, whether rock or grass or forest soil.

Nearly three years later, I write to him, asking what that's like. Within hours, he replies. The subject of the e-mail: "Stuff I don't feel."

The funny thing is that now that I'm in a wheelchair I crave the polished smooth surfaces of the hospital hallways and dread the woodchip riddled path at one of my local dog parks, as it's like riding over the bumpiest path ever, sending shock waves to my neck. I fantasize about the tactile feel of texture beneath my feet; stepping on lightly frozen puddles in the street and feeling the crunch and tinkle of the thin ice breaking. As a kid I simply couldn't walk past those lightly frozen puddles. I couldn't stop until every piece of thin ice was broken. Sand and grass are big ones that I miss. As a super avid skier I miss the whomp of landing on my butt in deep snow. I wrote a damn poem about the tactile memories I had of going from the basement changing room out to the lake at our cabin. I realized, only in retrospect, that I was fondly and bittersweetly recalling all the different feelings that took place underfoot; from the cold, smooth tiles in the bathroom, to the tightly woven comfort of the carpeted indoor stairs, to the sandy grit at the entryway, to the remote potential of slivers on the deck, to the smooth slide of the worn wood outdoor stairs, to the narrow gaps of the slotted wood deck boards, to the welcoming sandy bottom of the lake. It absolutely brings a tear to my eye when I think about how much sensory input, and ultimately emotion, passed through my feet in that short walk.

A few minutes later, a second e-mail shows up, titled "Stuff I feel." It reads,

I can't feel the earth underfoot per se, only the relative smoothness or roughness of the surface under my wheels as it translates to my neck and head, where I still have real feeling.

If I think of walking, I think of the four years Luna and I shared a house in Reno, where the foothill trails were a quick ten-minute drive away. At least once a day, we would head into these hills and hike for an hour or more, Luna running free, checking in from time to time as I walked in a kind of meditation, loop paths again and again, lost in thought but never lost. At dusk and after, we usually walked alone except for owls, or coyotes, or hawks that sometimes left bright red hunting tales in trailside snow. For Luna, these foothills were dog heaven. With rabbits to chase, water holes to swim, canyon walls to scale, she would come back to me with her feet mud-caked, coat smelling of sage, contented. On the drive home she would sit in the backseat, gazing out the window as though she had done everything she had come here to do.

I believe that intimacy can be learned and practiced. That we have a history that need not define us. That we have the opportunity — and maybe more important, the instinct — to cultivate our connections with the life around us.

These repeated walks became part of me. Each time we reached a familiar trailhead and set out again, I would sigh — I am back, returned to ground I have grown to know and love. Watching my friend charging up canyon walls or racing after rabbits? Taking in the delight she experienced while exploring? I could have lived like that forever.

On a gray Saturday noon, street wet from morning rain, I decide to take a walk. I start in my best orange-laced boots, but at block's end I take them off and head into the park toward the most beautiful maple, half its rich red leaves lying under its limbs. Wet grass, soft ground. The leaves cling to my skin in browns and brick reds, yellow-greens and burnt orange, my feet soon plastered. Lying under the maple, I see birds racing across the sky, the black wet trunk and branches backed by gray morning clouds. From heels and thighs to back and shoulder blades, I press against bare ground — thinking how comfortable it feels, my lower back relaxing — and close my eyes. I think of what Luna would be doing if she were here now. I think how, a century ago, a Dakota Sioux named Red Bird wrote, "My heart and I lie small upon the earth like a grain of throbbing sand."

One hundred fifty years ago, Henry David Thoreau asked, "*Where* are we?" and "*Who* are we?" Hasn't the answer always been the same? Rockets into space, the deepest ocean, we may visit these places but they will never be where we belong.

The word of Thoreau's I like best is this: "daily," as in "Daily to be shown matter, to come in contact with it."

Sacred ground is not beyond our reach, not somewhere else. While some grounds make the sacred more accessible, and in some grounds we are more clearly invited to connect, if the sacred is anywhere, it is at our feet. We are in contact with it every time we walk out the door.

When I walk outside today and look at the ground, I see a concrete sidewalk and a Kentucky bluegrass lawn, but that is only the beginning. I see the prairie that once filled the skies with wild birds, the fields of battle full of sacrifice, and those who have known those grounds all their life. I see six-foot-tall, dark-green corn, deep beds of black soil, and the tundra berries above the thawing Alaskan permafrost. I see the forest clearing where centimeters below parklike grass lie the bones and ashes of so many who loved this world and each other every bit as deeply as we try to. And I see, every time and always, the northern Minnesota woods where Luna lies buried and where, as I set her body into the ground that is more sacred to me than any other, the idea for this book began.

What I have seen since then is this: that while the paved and the hallowed are often so because of decisions made before we came along, the choice between hell or sacred is ours to make. Much of what I have seen tells me that by not making this choice—or not even being aware of it—we contribute to our world's slow drift toward a time of real hell. And if that's what we want, let's at least be honest. But if we want instead a world where our existence is based on the truth of knowing the connections that keep us alive—the bonds between us and others, between us and the natural world—then let us choose now. Let us walk out our door, look down, and find the sacred underfoot, wherever we are. Let us look to the ground, and know we are once and always home.

Acknowledgments

One of the most common questions any author receives is, How long did it take you to write your book? I never know quite what to say. I mean, for how long have I been thinking about the themes and ideas and questions that eventually took the form of this book? Though I usually end up saying "two or three years" I never feel quite right, as the real answer is probably something like "all my life." I feel similarly when it comes to thanking all the people who have helped me write this book. As any author will tell you, writing a book isn't something you do on your own. The following is my best attempt at remembering all those who have contributed to *The Ground Beneath Us*. If I have forgotten anyone here, I won't forget the next time we meet.

First, my deepest thanks to those people I visited and interviewed, who shared their time and insight with me, and whose stories are found on these pages. Without your generosity this book would not exist.

Next, I thank those many people whose names do not appear but who welcomed my visit and who were absolutely instrumental in helping me gather my story. These include Lin Jensen in California, author of the books *Deep Down Things* and *Pavement*, with whom I spent a wonderful afternoon;

Nienke Bakker of the Van Gogh museum in Amsterdam, who one day walked me past the long lines and gave me a personal tour; Curt Meine in Baraboo, Wisconsin, for time spent at the Aldo Leopold Center; Virgil Foote, the Sioux spiritual leader who invited me to his sweat lodge; Henry Way of James Madison University, who welcomed my questions about geography early in this book's creation; Scott Thayer at Legends Golf Club; and Fred Swanson, with whom I spent a day wandering among the owls and old growth at the H. J. Andrews Experimental Forest outside Corvallis, Oregon.

I would also like to thank the following people for taking time to talk with me about their work, to answer my questions, and to help guide my research and story: Ted Auch; Bonnie Harper-Lore; Kelsey Bankert; Jodi Johnson-Maynard at the University of Idaho; Kate Scow of the University of California, Davis; Fred Bahnson of Wake Forest University; Dennis Blanton of James Madison University; Mike Houck of the Urban Greenspaces Institute in Portland; Justin Long at the Minneapolis Park and Recreation Board; Scott Loss of Oklahoma State University; Janet Davis of Charlottesville; Kathleen Brown at Penn State University; David A. Robinson in Wales; Kelly Ramirez in the Netherlands; Christina Siebe and Silke Cram in Mexico City.

For their gracious hosting of me during my research travels I would like to thank Zuly Zabala León and Katherine Guio Reyes of the Banco de la República in Colombia. My thanks to the wonderful people of the Hay Festivals, who brought me to Xalapa, Mexico, and Cartagena, Colombia. These trips opened up a new world for me, and I am forever grateful. Ulla Remmer brought me to Vienna for an incredible event at the Naturhistorisches Museum, where after my reading we (audience included) went up onto the museum's rooftop to look down onto the city lights. Sabine Nikolay and her husband, Alexander, were incredible hosts on both my visits to Austria. In the Netherlands, Anna Christien Piebenga took me Wadlopen ("mud-walking") in the Wadden Sea. And in Mexico, the amazing Myriam Vidriales was my host in both Xalapa and Mexico City. One night, in the hacienda outside Xalapa owned by the wonderful Marisa Moolick, Myriam organized a dinner and reading attended by several highly esteemed writers and editors. After we opened the windows to the sights and sounds of November darkness, I read from *The End of Night* in English and Myriam read the same passages from the Spanish-language translation. It was an experience I will never forget, and I thank Myriam for her work on my behalf and for her friendship during my visits to Mexico and Colombia.

Closer to home, I would like to thank Julie Pfeiffer of Hollins University for many good conversations while the book was in its first stages. I thank Laura Greenleaf for her friendship and ideas, and for her work on behalf of darkness in Virginia. In Harrisonburg, my thanks to friends including Rob Alexander and Chip Brown, Karina Kline-Gabel and Kevin Gabel, Laurie Kutchins and Kevin Reynolds, Erica Cavanaugh and John Schaldach. My thanks as well to Alicia Horst, my friend and companion on many a Virginia adventure.

At James Madison University, I thank my wonderful colleagues in the English department, with special thanks to Dabney Bankert, Annette Federico, and Inman Majors. To David Jeffrey, dean of the College of Liberal Arts, my thanks for his steady support of my work. And to Shanil Virani, director of the John C. Wells Planetarium and indefatigable defender of dark skies and all things Canadian.

I owe a serious debt of gratitude to my former student

Katherine (Katie) McCombie, who with her research on my behalf went above and beyond what I could have imagined. What a pleasure it was to work with someone who so earnestly and thoughtfully took my all-too-often vague directions and returned pages of useful material.

My thanks to the Center for Energy, Environment, and Sustainability (CEES) at Wake Forest University for its support of my research travel, with a special thanks to its director, Miles Silman. I also want to share my gratitude for a Geospatial Technology Faculty Mini-Grant from James Madison University, which helped me learn more about new mapping technologies.

My ongoing thanks to Farley Chase of the Chase Literary Agency. There's always a cold Summit waiting if you get back to Minnesota. To Chris Jerome, my copy editor, thanks for catching and correcting and polishing up. I hope to see you again next time. And to all the folks at Little, Brown — thank you so much for your work on my behalf. Thanks especially to Carrie Neill. And above all to John Parsley — I give thanks every day that I have the good fortune to work with such a friendly, thoughtful, and insightful editor.

In Reno, I am thankful for continued friendships with Matt and Katie Menante Anderson, Jim Frost and Vanessa Belz, Dan Montero and Renee Aldrich. And to my friend Sudeep Chandra from the University of Nevada, Reno — here's to many more Indian buffets, beers on the patio, and conversations about changing the world.

To good friends near and far, including Scott Slovic, Eric Stottlemyer, Omaar Hena and Gretchen Stevens, Rajat Panwar, Cynthia Belmont, Erica Hannickel, Rachel Menke, Carly Johnson, Scott Dale and Andy Rusk, John and Laura Gibson, Kristin Tollefson, Andrew Comfort, Mike Macicak and Carmen Retzloff, Scott Dunn and Jill Richards, Robin Preble and Dan Hedlund, Heather McElhatton, Ingrid Erickson, Christina Robertson, and Emily Spiegleman.

I give thanks for continued conversation and adventure with Joshua Powell in Nashville, Randall Heath and Christine Keller in San Francisco, Andy and Tiffany Threatt Burelle in Albuquerque, and Marty Huenneke in Archlands. Special thanks to Douglas Haynes in Madison and to Thomas Becknell in St. Paul, both of whose friendship and advice have been invaluable in helping to shape this book. And to David Swirnoff in Minneapolis, a friend for thirty-five years and the first reader for most of these pages, my continued gratitude.

In the fall of 1998, I brought home an eight-week-old Brittany puppy and named her Luna. For fifteen years she shared my life, in New Mexico and Minnesota, Nevada and Wisconsin, North Carolina and Virginia. In August of 2013 we said goodbye. She was the best dog friend there ever could be, and our walks nearly every morning, evening, and night of her life probably brought me closer to writing a book about the ground than anything else I know. I miss her every day, and I am forever grateful for the time we shared.

It feels impossible to express in words the gratitude I feel for my parents, John and Judith Bogard. It's like giving thanks for life itself. But with this book I have tried.

And finally, to Caroline Hilk. On that first night, downstairs at Burch, I pretty much knew. A few weeks later, on a six-hour tour of Auschwitz during our first day in Europe together, I knew even more. With every Saturday morning, every conversation about what to write next, every minute cycling around the city, every laugh of surprise, every other countless little thing, I know again. Here's to walking this new ground together.

Notes

Introduction

This book's epigraph comes from Henry David Thoreau's description of his time on Mount Katahdin in *The Maine Woods*, published posthumously in 1864. Just over one hundred fifty years later, in the summer of 2016, President Barack Obama designated Katahdin Woods and Waters National Monument, 87,500 acres of the wilderness Thoreau wrote about in his book. "I looked with awe at the ground I trod on," he reported, "to see what the Powers had made there, the form and fashion and material of their work. This was that Earth of which we have heard, made out of Chaos and Old Night. Here was no man's garden, but the unhandselled globe. It was not lawn, nor pasture, nor mead, nor woodland, nor lea, nor arable, nor waste-land. It was the fresh and natural surface of the planet Earth."

According to the Environmental Protection Agency, the average American spends 93 percent of his or her life indoors. That includes 87 percent in buildings and 6 percent in cars. That also means that we spend, on average, only 7 percent of our life outside. See "The National Human Activity Pattern Survey" (NHAPS) in *Journal of Exposure Analysis and Environmental Epidemiology* 11 (2001): 231–252; 10.1038/sj.jea.7500165.

In the United States alone, pavements and other impervious surfaces cover more than 43,000 square miles — an area nearly the size of Ohio — according to research published in the June 15, 2004, issue of *Eos*, the newsletter of the American Geophysical Union.

For more on soil sealing, see "Assessment of Soil Sealing Management Responses, Strategies, and Targets Toward Ecologically Sustainable Urban Land Use Management," *Ambio* 43, no. 4 (May 2014): 530–541. See also "Soil Sealing" from the Global Soil Forum at globalsoilweek .org/wp-content/uploads/2014/11/GSW_factsheet_Sealing_en.pdf.

Especially as I began this book, the idea of pilgrimage was important. To walk the Gettysburg battleground, for example, where the First Minnesota's charge saved the second day; to see for myself Nebraska's Platte River with migrating sandhill cranes gathered tens of thousands strong—these were places I had always wanted to know. In *The Art of Pilgrimage*, Phil Cousineau writes that pilgrimage "means following in the footsteps of somebody or something we honor to pay homage." That is what often drew me to the places I went.

A dear friend of mine, Thomas Becknell, referred me to Cousineau's book, and while I did not directly focus on the idea of pilgrimage, it's remarkable to me now how much of what he says about pilgrimage rings true. For example, pilgrimage is, he writes, "transformative travel to sacred places." And how to get there? "The oldest practice is still the best. Take your soul for a stroll. Long walks, short walks, morning walks, evening walks…walking is the best way."

One gift of physical travel, of getting away from where you normally live, is that of perspective—you see your home anew. But ultimately most of us want to return. As Cousineau says, "The true pilgrimage is into the undiscovered land of your own imagination, which you could not have explored any other way than through these lands." This reminds me that Henry David Thoreau—who went to the Walden woods to gain perspective on life in his hometown—encouraged each of us to be "the Lewis and Clark of your own streams and rivers," and proclaimed with some pride that he had "traveled a good bit in Concord"—the town where he lived. I could have gone almost anywhere; what mattered is that I would go.

Manhattan

How do you grow four hundred trees and several stretches of lawn two hundred feet above the ground on the roof of a national memorial? At the National September 11 Memorial the turfgrass grows in beds eighteen inches deep, rooted in a sandy mix topped by three inches of topsoil. The swamp white oaks come in ten-by-ten crates, together needing forty thousand tons of soil. The tree roots will follow a dripline through tunnels, growing together beneath the concrete. Eventually, the trees will grow eighty feet tall, coming together to form a canopy. Just as the grass invites but is off limits, so this memorial is nature but not really. Even if you want to invite millions of visitors under the healing shade of growing trees, you still need to keep their boots and shoes on concrete.

While the oldest "shoes" found thus far have been dated back thousands of centuries, in eras such as that of ancient Greece, whole societies were essentially shoeless. The original Olympic athletes, for example, participated without shoes (or clothes, for that matter), and even the society's gods and heroes were portrayed without shoes. And it's crazy to imagine that Alexander the Great created his empire with armies of barefoot soldiers.

For most of us, walking is easy to take for granted, but as one study notes, walking is a "complex dynamic task that requires the generation of whole-body angular momentum to maintain dynamic balance while performing a wide range of locomotor subtasks such as providing body support, forward propulsion and accelerating the leg into swing." See "Muscle Contributions to Whole-Body Sagittal Plane Angular Momentum during Walking" in the *Journal of Biomechanics* 44, no. 1 (January 4, 2011): 6–12. Or, in other words, there's a whole lot going on in our bodies as we walk down the street. Our muscles "accelerate body segments and generate ground reaction forces that alter angular momentum. In addition, gravity contributes to whole-body angular momentum through its contribution of the ground reaction forces." As have bears and great apes, we humans are among the few animals to have evolved walking on our heels first, then rolling onto the ball of the foot and our toes. In fact, this heel-first approach is 53 percent more efficient than walking on the balls of our feet, and a whopping 83 percent more efficient than walking on our toes.

Our human ancestors began walking some three to seven million years ago, with *Homo sapiens* taking foot 200,000 years in the past, and we have the bones, muscles, and breathing to show for it. Fossil bones show the evolution of a gradual ability to walk upright—thigh bones angled to support the body's weight, knee bones made to absorb the stress of briefly standing on one leg as we walk, the curve of our spine acting as a shock absorber while connecting underneath our skull to hold our head upright. By comparison, a chimpanzee's spine attaches at the back of its head, keeping the head at an angle; its thigh bones and knees can't support walking upright for long. Scientists suspect that we evolved to walk upright in order to better access food—reach for higher fruits, chase mammals, travel to new grounds. In other words, not so different from our lives today.

Find the article "To Age Well, Walk" at http://well.blogs.nytimes.com/2014/05/27/to-age-well-walk/?_r=0. Here, Gretchen Reynolds reports the results of "one of the largest and longest-running studies of its kind," published in the *Journal of the American Medical Association*, explaining that "regular exercise, including walking, significantly reduces the chance that a frail older person will become physically disabled." Find Tom Vanderbilt's fascinating article "The Crisis in American Walking" in Slate at http://www.slate.com/articles/life/walking/2012/04/why_don_t_ameri cans_walk_more_the_crisis_of_pedestrianism_.html. Says Vanderbilt, "America is a country that has forgotten how to walk."

In *Beyond Geography: The Western Spirit Against the Wilderness* (1982), Frederick W. Turner described a world none of us will ever know: "To those who followed Columbus and Cortes the New World truly seemed incredible, not only because of what civilization had made of the Old World but because of the natural endowments of the one they now began to enter. The land often announced itself with a heavy scent miles out into the ocean....Giovanni di Verrazano in 1524 smelled the cedars of the East Coast a hundred leagues out...(and) ships running farther up the coast occasionally swam through large beds of floating flowers."

For more on John Randel Jr. and his mapping of Manhattan, see "The Grid at 200: Lines That Shaped Manhattan," at http://www .nytimes.com/2012/01/03/arts/design/manhattan-street-grid-at -museum-of-city-of-new-york.html. See also "No Hero in 1811, Street Grid's Father Was Showered With Produce, Not Praise," at http://his torynewsnetwork.org/article/137854. Marguerite Holloway's book, *The Measure of Manhattan: The Tumultuous Career and Surprising Legacy of John Randel Jr.*, was published in 2013.

Like so many others, I "love maps." When I say that, I don't mean the highly functional map on my iPhone, the map I use nearly every day. I mean paper maps, the kind you can fold out on the table before you, lean over, and study with joy. I'm certainly old enough to remember when these were the only kinds of maps we had, and at least for me, when I say I love maps, one thing I'm saying is that I love the feeling of curiosity that comes over me as I view them, a sense of traveling while barely leaving my seat. I so often have a sense of wanting to go to places on the map. Stephen S. Hall writes, "The most important thing a map shows, if we pause to look at it long enough, if we travel upon it widely enough, if we think about it hard enough, is all the things we still do not know."

My thanks to Shanil Virani, director of the John C. Wells Plane- tarium at James Madison University, for helping me to understand the reasons we aren't flung from the Earth's surface as it spins at a thou- sand miles an hour.

As you might expect, NASA offers a wealth of information on gravity, or rather the lack of it. One example, "The Beating Heart, Minus Gravity," reports that astronauts who appear to be floating in space are actually in a kind of free fall, being pulled by gravity toward Earth. And lucky them, for if this weren't the case, they "would travel in a straight line away from the Earth." For more on gravity, see Brian Clegg's book *Gravity: How the Weakest Force in the Universe Shaped Our Lives* (St. Martin's, 2012).

Perhaps the thing that is most impressive about Eric Sanderson is his vision, both of the past and the future. For example, when Sanderson looks four hundred years into the future, he likes what he sees. A city, a megacity, in this case New York City, with more inhabitants than now, with streams running alongside streets, streets with few cars, cars without drivers. He sees local produce from countryside close by, the surrounding New York and New Jersey farmland reborn, returned to nineteenth-century roots when it was the largest farming market in the country. His is the kind of vision that can inspire, that while realistic about our coming struggles can imagine us thriving beyond them. He can see not only the world that we are losing but the one we might create. For more, see his wonderful *Terra Nova: The New World After Oil, Cars, and Suburbs* (Henry Abrams, 2013). See also his work on "The Nature of Cities" at http://www.thenatureofcities.com/author/ericsanderson/.

Originally the New York Zoological Society, Eric Sanderson's employer the Wildlife Conservation Society "works to conserve more than two million square miles of wild places around the world." Find them at https://www.wcs.org. Find the World Wildlife Fund study that Sanderson quotes, "Living Planet Report 2014," at http://www.worldwildlife.org/pages/living-planet-report-2014. The dire news: "Population sizes of vertebrate species—mammals, birds, reptiles, amphibians, and fish—have declined by 52 percent over the last 40 years."

Learn more about the maps created by Sanderson and his colleagues at http://bioscience.oxfordjournals.org/content/52/10/891.full. For better or for worse, they report, "the influence of human beings on the planet has become so pervasive that it is hard to find adults in any country who have not seen the environment around them reduced in natural values during their lifetimes." Sanderson and his colleagues call land transformation "the single greatest threat to biological diversity." Nonetheless, they report, "This phenomenon and its implications are not fully appreciated by the larger human community, which does not recognize them in its economic systems or in most of its political decisions."

Among their effects, the maps created by Sanderson and his colleagues could especially address two different sets of people. The first is people who have no idea—people who think that if you get out of the city, or the state, or the country, there is plenty of "wild" left. Sanderson's maps show that this simply isn't true. But the maps are also important for someone who has heard enough about what's happening in the world to wonder, What's the use of trying to make a difference? As I learned over the course of my journey, there is no shortage of compelling reasons to be trying as hard as possible right now.

Another set of maps shows that "urbanization, agriculture, and energy could gobble up 20 percent of the world's remaining natural land by 2050." The authors argue that "with development increasingly encroaching into more remote and previously undisturbed areas, it is critical that international corporations, governments and conservation organizations collaborate to reduce and minimize potential future impacts on remaining habitats." Find the study "A World at Risk: Aggregating Development Trends to Forecast Global Habitat Conversion," at http://journals.plos.org/plosone/article?id=10.1371/journal.pone.0138334#pone-0138334-g004.

A recent study published in the *Proceedings of the National Academy of Sciences* argues that "'Pristine landscapes simply do not exist and, in most cases, have not existed for millennia": see http://phys.org/news/2016-06-pristine-landscapes-havent-thousands-years.html. For some, this would seem to undermine efforts to conserve wild places. As a *Washington Post* article began, "Implicit in much, if not all, modern environmental sentiment is the idea that the natural world has been despoiled by humans—and if we could just leave it alone, things would get better": visit https://www.washingtonpost.com/news/energy-environment/wp/2016/06/06/theres-basically-no-landscape-on-earth-that-hasnt-been-altered-by-humans-scientists-say/?utm_term=.8d72c45b2d5a. But this is a shallow understanding of "modern environmental sentiment." Far more common is the understanding that areas don't have to be "pristine" or "wild"—which anyway are human conceits—to be worthy of conservation. Most environmental thinkers understand that especially if we include in the human footprint such human-caused influences as climate change and chemical pollution, there are no such things as "pristine" or completely "wild" areas left on Earth. But so what? There are still places of compelling beauty and biodiversity that deserve our greatest efforts and care.

Find Maya Lin's powerful memorial—one in which each of us is invited to participate—at http://www.whatismissing.net. The memorial's goals? To "raise awareness of what we are losing and to show you what you can do to help."

For a wonderful image of the built environment below Manhattan streets, see *National Geographic*'s "New York Underground" at http://www.nationalgeographic.com/features/97/nyunderground/.

Mexico City

In 1864, the same year Thoreau's *The Maine Woods* was published, and a year after the battle of Gettysburg, Jules Verne published *Journey to the*

Centre of the Earth. The story of three men traveling under Iceland's volcanoes, the novel is nearly as fascinating today as it must have been at its publication. "Words of human language are inadequate to describe the discoveries of one who ventures into the deep abysses of the Earth," wrote Verne.

Of course, no human has actually journeyed to the Earth's center, and the trip is impossible for a number of reasons, not the least of which the heat would melt any human who tried. Anyway, to get there would first require cutting through 20 to 45 miles of crust, followed by 1,800 miles of mantle, and then through the 10,000-degree liquid-iron core, not to mention overcoming the pressure exerted by 6.6 sextillion tons of rock pressing inward.

This doesn't mean that humans aren't continuing to drill into the Earth. For one fascinating story about doing just this, listen to NPR's "Drilling to the Mantle of the Earth," at http://www.npr.org/2011/03/25/134855888/Drilling-To-The-Mantle-Of-The-Earth.

Along with the growth in the number of megacities comes the growth of megaregions, where several megacities are joined together. The America 2050 initiative, reported at http://www.america2050.org/content/megaregions.html, shows the United States as having several megaregions by 2050, and not only the Eastern Seaboard or Los Angeles, but also areas including the Front Range of Colorado, the Seattle–Portland corridor, and the Atlantic–Charlotte region. The challenge for conservationists is how to include healthy nature (and access to nature for humans) amid all the development.

To learn more about all things concrete, see Robert Courland's *Concrete Planet: The Strange and Fascinating Story of the World's Most Common Man-Made Material* (Prometheus, 2011). Courland told me simply, "Concrete dominates the human environment. Nothing is as representative of modern and postmodern society as steel reinforced concrete." As the *New York Times*'s Elisabeth Rosenthal writes, "cement is literally the glue of progress." One of the main reasons to pay attention to our paving of the world and our use of concrete is, as Courland told me, "making concrete cement is particularly a dirty business." In fact, the production of concrete cement ranks second only to automobiles and coal-fired power plants in generating greenhouse gases. "The cement industry admits to generating 10 percent of all the CO_2 produced," Courland explained, "but this is a conservative estimate...it's probably at least 15 percent." While cement manufacturers have spent millions in an attempt to make their industry more "green," the skyrocketing demand for concrete cement around the world has essentially canceled out those improvements.

For more information on how Mexico City is sinking, see http://www.homelandsecuritynewswire.com/mexico-citys-sinking-worsening), as well as "Why Is Mexico City Sinking?" at https://www.theguardian.com/science/2004/may/06/thisweekssciencequestions.

"They are about to go extinct," says biologist Sandra Balderas Arias of the axolotl, in an article titled "Mythic Salamander Faces Crucial Test: Survival in the Wild" at http://www.nytimes.com/2012/10/31/world/americas/struggle-of-axolotls-mexicos-mythical-salamander.html. For more, see "In Mexico, the ajolote's fate lies in troubled waters" at http://articles.latimes.com/2012/oct/01/world/la-fg-mexico-magic-salamander-20121002.

One of Mexico's greatest gifts to the visitor is its museums. See especially the National Museum of Anthropology, where the nation's incredibly rich human history is on display. In the section covering the Aztecs, one sculpture description reads, "This large container for the food of the gods, mainly blood and human hearts."

Before my visit to Mexico City, I spoke with Dennis Blanton, an archeologist in Virginia, about his profession. He told me he sees as his goal to find the human story in the ground, especially from those times in history when people weren't telling that story in writing. I asked, Why should we do that? "The most straightforward answer is that we build our identities around heritage," he had said. "We're not made out of a vacuum of thoughts; we are products of the heritage and the past. It shapes us, and in less than obvious ways. And the fundamental issues that confront us now? They're not new ones. If you want to understand present circumstances, even yourself, you have to reflect back."

In his seminal book *Life and Death in the Templo Mayor*, Eduardo Matos Moctezuma describes the "unique opportunity to break through the thick barrier of concrete that covers the city of Tenochtitlán and, archeologically, to peer through the window of time and recover a time gone by." After spending time with Matos Moctezuma, it's clear to me that our modern manner of living separated from the natural cycle of life and death would have completely confused the Aztecs. "You must keep the cycle going," I can imagine them saying. "You must recognize that in order for us to live, something must die." Returning from the fifteenth century, as I stepped from the site, I found the Aztec values far less crazy than I might have before my visit. The carving out of human hearts may still repulse me, but not the values behind the ritual. Not the notion that ensuring that the creation that sustains us continues is largely in our hands.

The graphic from *National Geographic*, as well as a wonderful article on the Aztecs, can be found at http://ngm.nationalgeographic.com/2010/11/greatest-aztec/draper-text.

London

The epigraph comes from Wendell Berry's brilliant book *The Unsettling of America: Culture and Agriculture* (Sierra Club, 1977).

London's Crossrail project has garnered plenty of deserved publicity. The intricacy of placing a railroad underneath one of the world's oldest cities and the archeological discoveries that have resulted have fascinated the world. One wonderful source is the *National Geographic* article "London's Big Dig Reveals Amazing Layers of History," from its February 2016 issue: visit http://ngm.nationalgeographic.com/2016/02/artifacts-found-under-london-archaeology-text. *The New Yorker*'s article ("Bedlam's Big Dig") includes this observation: "There is a lot of death here in Liverpool Street," Jay Carver, Crossrail's lead archeologist, said as we looked down from a platform over the site. "A lot of dead people,": it's at http://www.newyorker.com/tech/elements/london-crossrail-bedlam-big-dig. Other articles include such cleverly descriptive titles as "London Crossrail Dig Hits Beheaded Romans" (*Forbes*), "The Monster Tunneling Under London Streets" (BBC), and "Cremated human bones in pot found in Crossrail dig suggest gruesome ritual" (*The Guardian*).

The treasures found during the Crossrail dig continue London's long underground history. Peter Ackroyd's *London Under: The Secret History Beneath the Streets* (Doubleday, 2011) is a great place to learn more. Incredibly, the history lies immediately below your feet. As you walk in the city, Ackroyd explains, "You are also treading on the city of the past, all of its history from the prehistoric settlers to the present day packed within 24 feet of earthen fabric." In terms of dead people, echoing Jay Carver above, Ackroyd reports that "from Roman London alone there issued a million corpses."

One of the most famous corpses to be found in recent years—though not in London but in Leicester—was that of King Richard III, whose remains were found buried under parking lot asphalt. For more on this mind-boggling story, see "DNA Confirms: Here Lieth Richard III Under Yon Parking Lot" at http://news.nationalgeographic.com/news/2014/12/141202-richard-iii-genes-shakespeare-science/.

Speaking of Roman London, visitors to the modern city can purchase a map of Londinium made by the Museum of London that superimposes the historical city over the much larger modern city boundaries. Anyone visiting the modern city would do well to spend time at the Museum of London, a sometimes forgotten attraction that holds a wealth of information and story about one of the world's most fascinating cities.

To find out more about Graham Rook's work, see the articles "Regulation of the immune system by biodiversity from the natural

environment: An ecosystem service essential to health" at http://www
.ncbi.nlm.nih.gov/pubmed/24154724, and "Microbial 'old friends,' immu-
noregulation and socioeconomic status" at http://www.ncbi.nlm.nih
.gov/pubmed/24401109. "Epidemiological studies suggest that living
close to the natural environment is associated with long-term health
benefits," begins the first paper. Or, as Rook told me, "It was just very
striking that you will more likely drop dead in that five-year period (of
a UK study)...if you did not live close to green space."

Michael Pollan's article "Some of My Best Friends Are Germs"
appeared in the *New York Times Magazine* on May 15, 2013.

The titles of recent articles point to an increased awareness of the
value of dirt for human health. "Is Dirt the New Prozac?" (*Discover*),
"How the 'Dirt Cure' Can Make for Healthier Families" (*New York
Times*), and "Healthy Soil Microbes, Healthy People" (*The Atlantic*) are
just a few. When it comes to public health, this is a change. "Enter the
terms 'soil' and 'health' into a PubMed database," reports Daphne
Miller, MD, in an article titled "The Surprising Healing Qualities...of
Dirt," "and the top search results portray soil as a risky substance, filled
with pathogenic yeasts, antibiotic-resistant bacteria, radon, heavy met-
als, and pesticides." But move beyond these reports, Miller says, and
"you will discover a small, but growing, collection of research that
paints soil in a very different light"; see http://www.yesmagazine.org/
issues/how-to-eat-like-our-lives-depend-on-it/how-dirt-heals-us.

For more on the link between soil and human health, an excellent
place to start is the work of Dr. Eric C. Brevik. See "The past, present,
and future of soils and human health studies" at http://www.soil
-journal.net/1/35/2015/soil-1-35-2015.pdf, as well as his book *Soils and
Human Health* (Taylor & Francis, 2012).

As more and more of us live in cities, contact with the natural
ground will become increasingly vital. Green spaces and parks, trees
and water—an ever-growing list of studies support the idea that we
are healthier people when we have contact with nature. "How green
cities are better for us physically and psychologically" (*Globe and Mail*),
"Urban green space delivers big happiness boost" (*Conservation*), and
"Why living around nature could make you live longer" (*Washington
Post*) are typical of recent headlines making this case. If we accept this
as fact, the implications for the ways we develop our cities are many. If
we want to be healthy, these studies seem to argue, we cannot continue
to pave the world. Unfortunately, in many cities around the world, living in urban areas means being cut off from nature. "Global Urbanization and the Separation of Humans from Nature" reports that "within
cities worldwide, most residents are concentrated in neighborhoods of

impoverished biodiversity." Too often, those most affected are children. For more information on the story of the *Oxford Junior Dictionary* removing nature-oriented words, see "Panic at the Dictionary" at http://www.newyorker.com/books/page-turner/panic-dictionary).

While it has yet to gain traction in the medical community, the idea of "grounding" or "earthing" for human health has an increasing number of supporters. "Multi-disciplinary research has revealed that electrically conductive contact of the human body with the surface of the Earth produces intriguing effects on physiology and health," reports one recent article. See "The effects of grounding (earthing) on inflammation, the immune response, wound healing, and prevention and treatment of chronic inflammatory and autoimmune diseases" at http://www.ncbi.nlm.nih.gov/pmc/articles/PMC4378297/.

For more on E. O. Wilson's theory of biophilia, see his book *Biophilia: The Human Bond with Other Species* (Harvard, 1984). Among Wilson's many other books are *Consilience* (1999), *Letters to a Young Scientist* (2014), *The Meaning of Human Existence* (2014), and *Half-Earth: Our Planet's Fight for Life* (2015). A scientist with the ability to translate data into story, Wilson—well into his eighties—continues to do all he can to protect and conserve the world's biodiversity. Without this life, he warns, we risk entering a new age he calls "Eremocene, the Age of Loneliness": see http://www.economist.com/news/21589083-man-must -do-more-preserve-rest-life-earth-warns-edward-o-wilson-professor -emeritus. Wilson is well aware of the life at our feet, writing in his book *The Creation*, "Each cubic meter of soil and humus within it is a world swarming with hundreds of thousands of such creatures, representing hundreds of species. In one gram of soil, less than a handful, live on the order of ten billion bacteria belonging to as many as six thousand species."

For helping me understand how we might design our cities with biophilia in mind, my thanks go to Dr. Timothy Beatley of the University of Virginia. "A biophilic city is a city abundant with nature," he writes in his *Biophilic Cities* (Island, 2010), "a city that looks for opportunities to repair and restore and creatively insert nature wherever it can." Find out more about biophilic cities at biophiliccities.org. Among other things, Beatley writes, "a biophilic city is an outdoor city, a city that makes walking and strolling and daily exposure to the outside elements and weather possible and a priority." Appropriately, my conversations with Tim took place while we strolled the UVA campus, one designed by the man Beatley calls our first "biophilic president," Thomas Jefferson.

For more on the Campaign to Protect Rural England, visit http:// www.cpre.org.uk. One of the unfortunate trends that CPRE is trying to slow is that of the paving of front yards in England, mostly to provide parking space for additional automobiles. Called "front gardens" in the UK, these patches of once-natural ground have been paved at an incredible rate—in London, for example, the number paved was up 36 percent in just a decade. See http://www.independent.co.uk/news/uk/home-news/three-times-as-many-front-gardens-completely-paved-as-a-decade-ago-says-royal-horticultural-society-10256660.html.

Northern Virginia

Published in 1993, James Howard Kunstler's *The Geography of Nowhere: The Rise and Decline of America's Man-Made Landscape* (Touchstone) remains on point nearly twenty-five years later. The entire book is Kunstler's attempt to wake us from our lack of attention to the environment around us. "Americans evince a striking complacency when it comes to their everyday environment and the growing calamity that it represents," he writes.

The Cedar Creek Battlefield is easily found near the intersection of Interstates 81 and 66 in northern Virginia. If you go, make sure to visit the small museum at Hupp's Hill Civil War Park in Strasburg, where Mike Kehoe's many archeological finds fill the display cases. For more information, see http://ccbf.us.

Find out more about the map of roads with every other feature removed at http://io9.gizmodo.com/a-map-of-u-s-roads-and-nothing-else-483183413. The high-resolution version of the map is especially striking. See also "These crazy maps show just how much ground roads cover" at http://grist.org/living/these-crazy-maps-show-just-how-much-ground-roads-cover/.

The study by Nick Haddad and others, "Habitat fragmentation and its lasting impact on Earth's ecosystems," can be found at http://advances.sciencemag.org/content/1/2/e1500052. For an excellent article on this study and on the effect of roads in general, see "What Roads Have Wrought" by Michelle Nijhuis in *The New Yorker* at http://www.newyorker.com/tech/elements/roads-habitat-fragmentation. The information that "more than 25 million kilometres of new roads will be built worldwide by 2050" comes from the article "Study shows where on the planet new roads should and should not go" at http://www.cam.ac.uk/research/news/study-shows-where-on-the-planet-new-roads-should-and-should-not-go, which details a new report published in the journal *Nature*, "A global strategy for road building" at http://www.nature.com/nature/journal/v513/n7517/abs/nature13717.html.

Despite studies such as these, roads continue to be built in places where they will almost certainly do significant ecological harm. The American ecologist Aldo Leopold saw this in the 1940s as he was writing *A Sand County Almanac*. The line from that book, "To build a road is so much simpler than to think of what the country really needs," remains—as does so much of Leopold's work—prescient for our day. For example, a paved road is proposed that would cut directly through Tanzania's Serengeti National Park. Despite the fact that a road could be built that would meet the same goals but avoid the park by staying south of its border—the construction of which would be supported by the German federal government—the Tanzanian government continues to threaten the most disruptive option. For anyone interested in helping stop what scientists say would be an ecological catastrophe, visit Serengetiwatch .org. "If we can't save the Serengeti," they ask, "what can we save?"

One bright spot comes in the continued success of "The Roadless Rule," originally declared by President Bill Clinton in 2001, a rule that protects more than fifty-eight million square miles of roadless areas in US national forests. The battle over this rule, its implementation and maintenance, is explained in "The Nine Lives of the Roadless Rule," by Harvard's Richard Lazarus at environment.law.harvard.edu/ wp-content/.../12/Lazarus_Forum_2015_Nov-Dec.pdf.

Much has been written about the costs of sprawl to human and environmental health. Among the most helpful to understanding these costs are "Vanishing Open Spaces: How an Exploding US Population Is Devouring the Land that Feeds and Nourishes Us" by NumbersUSA, at https://www.numbersusa.com/resource-download/vanishing-open -spaces; "Paving Paradise: Sprawl and the Environment" at nrdc.org/cities/ smartgrowth/rpave.asp, from the Natural Resources Defense Council; and "Paving Paradise: The Peril of Impervious Surfaces" by Lance Frazer at http://www.ncbi.nlm.nih.gov/pmc/articles/PMC1257665/. The latter includes the information that "paved surfaces are quite possibly the most ubiquitous structures created by humans."

In addition to their value for our physical, mental, and spiritual health (not to mention their value for wildlife, plants, and soil), one of the most compelling reasons for preserving open spaces is their monetary value. Factors such as their effect on property values, their value in terms of environmental services such as groundwater rejuvenation, and—especially—avoiding the inevitable costs of sprawl, mean hundreds of millions of dollars for surrounding communities. See, for example, http:// thecostofsprawl.com, and "How Much Sprawl Costs America" at http:// www.citylab.com/housing/2015/03/how-much-sprawl-costs -america/388481/. The estimate? One trillion dollars a year. That these open spaces are often sacrificed in a way that benefits developers and investors rather than local communities is something we would be wise to better understand. Unfortunately, explains the report from NumbersUSA, "Sprawl and the loss of open spaces themselves don't get nearly the attention from the news media, politicians and national public interest groups that they did a decade ago. But the losses have not stopped."

One of the most damaging effects of paving over natural ground is creation of impervious surfaces. The transformation of previously natural ground into impervious surfaces such as roads, parking lots, and roofs has disrupted the natural hydrologic cycle of rainfall soaking into the ground. As a result, "stormwater runoff" picks up pollutants that poison streams, rivers, and wetlands, and keeps it from recharging aquifers. Runoff pollution is now the leading threat to the nation's water quality, affecting about 40 percent of surveyed rivers, lakes, and estuaries. Bodies of water all around the country, from the Great Lakes to the Everglades to the rivers of the Pacific Northwest, are being damaged by runoff from paved impervious surfaces.

For an excellent summary of Eran Ben-Joseph's study of parking lots, see "Paved, but Still Alive" by Michael Kimmelman of the *New York Times* at http://www.nytimes.com/2012/01/08/arts/design/taking -parking-lots-seriously-as-public-spaces.html?_r=0. Kimmelman reports that in his "Rethinking a Lot," Ben-Joseph points out that "in some US cities, parking lots cover more than a third of the land area, becoming the single most salient landscape feature of our built environment."

For a wonderful article on the pressures from development faced by US Civil War Battlefields, see Adam Goodheart's article "US Civil War battlefields see new conflict" at http://ngm.nationalgeographic .com/ngm/0504/feature5/. Goodheart tells the story of "a quickly vanishing America of small farms and crossroads villages and a newer landscape of megamalls and sprawling McMansions." Much of the Civil War landscape, he writes, "has been obliterated, often by developments whose names give hollow echo to the Civil War's guns — Artillery Ridge, Lee's Parke." Against such reckless development works the Civil War Trust. Find out more about their efforts at http:// www.civilwar.org.

While I knew the general history of the US Civil War in the Shenandoah Valley area (to which I moved in 2012), I had no idea of the depth and breadth of this history. One of the first to clue me in to this was Laura Greenleaf, a friend from the International Dark-Sky Association, who grew up in rural northwest Virginia and told me the story of the day she dug the grave to bury her childhood dog in her backyard and struck a Confederate belt buckle. My thanks to Laura for helping me understand how at the heart of, as she says, "the suburbanization of Virginia...is a loss of a sense of place, of place identity," and how because during the war "the scale of the action—the size of troop movement and encampments—were such, really no ground was untouched in the valley."

Gettysburg

The chapter's epigraph comes from Abraham Lincoln's Lyceum Address, delivered in Springfield, Illinois, on January 27, 1838. In the speech, the future president warned against the dangers of slavery in the United States.

For more on Lincoln's Gettysburg Address twenty-five years later, and the idea of consecrating the battlefield as "hallowed ground," see Jeffrey B. Roth's "Consecrating Hallowed Ground" from the *New York Times* at http://opinionator.blogs.nytimes.com/2013/11/20/consecrating-hallowed-ground/?_r=0. It's nearly impossible for the modern visitor to the battlefield to understand what it would have been like on November 19, 1863. Roth writes that "for months, the smell of decaying bodies saturated large portions of the 50 square miles around Gettysburg. Bodies, body parts, dead horses and the remains of military equipment littered the battlefield." When Lincoln toured the battlefield, Roth reports, "the skeletal remains of the rib cages of horses remained visible on portions of the field."

The authoritative book on the First Minnesota's heroics during the battle is Brian Leehan's *Pale Horse at Plum Run: The First Minnesota at Gettysburg* (Minnesota Historical Society, 2002). Originally from Minnesota himself, world-renowned historian James M. McPherson's *Hallowed Ground: A Walk at Gettysburg* (Crown, 2003) is a short and pleasant journey around the battlefield. Drew Gilpin Faust's best-selling *This Republic of Suffering: Death and the American Civil War* (Vintage, 2009) details how utterly unprepared Americans—including those in the two armies—were for the war's carnage.

On the third day of the Gettysburg battle, the forty-seven remaining soldiers from Minnesota took part in defeating Pickett's Charge. It was a battle in which the First Minnesota captured the battle flag of the Twenty-Eighth regiment from Virginia. Since then, the flag has been kept at the Minnesota capitol in St. Paul, and overtures from Virginia for its return—the latest coming in 2013—have been politely rebuffed. For more, see http://blogs.mprnews.org/statewide/2013/06/no-virginia-there-will-be-no-battle-flag-for-the-gettysburg-anniversary.

One of the questions Park Service employees at Gettysburg are sometimes asked is whether any of the trees on the battlefield today were "witness trees," if they were present at the battle in 1863. When I asked John Commito about this, he told me, "I don't believe there are any witness trees still on the battlefield. In my time out on the battlefield with rangers in various places, they'd say, No, you can look all around here, even though some of these trees are big, none of them are witness trees." But then he added, "It may be possible that there are witness trees that they're just not telling people about."

Kent Gramm's excellent books include *November: Lincoln's Elegy at Gettysburg* (Indiana University, 2001), *Gettysburg: This Hallowed Ground* (Tide-Mark, 2004), and *Sharpsburg: A Civil War Narrative* (Resource, 2015).

Bishopstone

Find Wendell Berry's wonderful essay "The Pleasures of Eating" at https://www.ecoliteracy.org/article/wendell-berry-pleasures-eating. It's said that a single sentence from Berry's essay inspired Michael Pollan's books about food and eating. See "The Wendell Berry Sentence That Inspired Michael Pollan's Obsession with Food" at http://www.theatlantic.com/entertainment/archive/2013/04/the-wendell-berry-sentence-that-inspired-michael-pollans-food-obsession/275209/. There are so many good sentences in this single essay that it's hard to choose those to quote, but here are a few more: "The food industrialists have by now persuaded millions of consumers to prefer food that is already prepared. They will grow, deliver, and cook your food for you and (just like your mother) beg you to eat it. That they do not yet offer to insert it, prechewed, into our mouth is only because they have found no profitable way to do so."

To find out everything you would want to know about Helen Browning, her work, her farm, visiting her farm, and buying her food, visit http://www.helenbrowningsorganic.co.uk. The farm now features a bed-and-breakfast, and the food at The Royal Oak is phenomenal. For more on Britain's Soil Association, visit www.soilassociation.org.

One morning during my visit to Helen's farm, fueled by a "proper English breakfast" from The Royal Oak featuring fried eggs, tomatoes, and bacon and sausage from Helen's pigs, I decided to hike along the Ridgeway, Britain's oldest road, which crosses close to Bishopstone. Now recognized as a National Trail some eighty-seven miles long, the Ridgeway has been a road for more than five thousand years. Helen told me that as a child she knew it simply as the "Roman Road." And indeed, the Romans used the Ridgeway, but so did scores of peoples before them, and scores afterward. The section I walked is basically a two-track gravel road about ten feet wide. No motor vehicles are allowed, just me and the horses carrying riders from out of the past. They came around corners, clip-clopping their way as though we were sharing a moment from two hundred years ago. "Morning," one said, but that was about it.

I walked to the Bronze Age Uffington White Horse, a massive pre-historic horse figure carved into the side of the ridge some two thousand to three thousand years ago. I could have sworn that Helen had described the horse as a "sculpture," and so as I was making my approach I was looking around for a big white horse carved of chalk, the kind of thing that might be rearing up outside Denver's airport. Streams of English visitors were passing me, teenagers in packs, families pushing strollers and pulling grandparents, everyone blown about by a blustery wind. "Is the horse around here?" I finally asked a passerby. "Sure is, mate," as he pointed to the other side of the ridge. I wandered over—I obviously hadn't yet remembered the cover of my 1986 XTC *English Settlement* album, or I would have known what it was I was supposed to see. And even when I did see it, I didn't. Because the Uffington White Horse "sculpture" is carved into the ground, and it's huge, up close a white chalk pit (the back of the horse? the legs?). I love that this is a sculpture in the ground—of the ground—and that it has lasted. The Uffington White Horse—it's like a tattoo on the earth, a mark that says we were here, once upon a time.

To get a quick sense of the incredible biodiversity in soil, visit http://www.fao.org/soils-portal/soil-biodiversity/facts-and-figures/en/. The web pages for the Food and Agriculture Organization of the United Nations are, in general, a great place to start learning more about soil.

To find out more about the experiment carried out in the soil of New York City's Central Park by Noah Fierer, Diana Wall, and others, see "Beneath Central Park, a Teeming Universe" at http://cityroom.blogs.nytimes.com/2014/10/02/beneath-central-park-a-teeming-universe/. Among the article's opening lines: "Biologists found a staggering 167,000 kinds of microbes living in the park's soil—the vast majority of them never before documented."

Dr. Diana Wall was an immense support for me while writing this book, spending time with me during my visit to her Colorado State University offices and labs, talking by phone, and visiting my classes during her visit to James Madison University. Among her achievements is leading the creation of the worldwide Global Soil Biodiversity

Initiative and the world's first Global Soil Biodiversity Atlas. Find out more about her work at http://wp.natsci.colostate.edu/walllab/people/dr-diana-h-wall/.

For more about the GSBI, visit https://globalsoilbiodiversity.org/node/271, and see "The Hidden World Under Our Feet" by Jim Robbins at http://www.nytimes.com/2013/05/12/opinion/sunday/the-hidden-world-of-soil-under-our-feet.html.

Increasing numbers of articles and studies report on the priceless value of microbes and microbiomes. For a sample, try "What lies beneath" at http://www.nature.com/news/2008/081008/full/455724a.html; "Is Climate Change Putting World's Microbiomes at Risk?" at http://e360.yale.edu/feature/is_climate_change_putting_world_microbiomes_at_risk/2977/; and "Scientists Urge National Initiative on Microbiomes" at http://www.nytimes.com/2015/10/29/science/national-initiative-microbes-and-microbiomes.html.

While there is no shortage of information on the brutal treatment of pigs in the US's industrial agriculture system, specifically in the confined animal feeding operations, for my money the most powerful and persuasive is Jonathan Safran Foer's remarkable *Eating Animals* (Little, Brown, 2009). I have no doubt ruined scores of students' Thanksgiving holidays after assigning the book during fall semester. After reading Foer's work, it's very hard not to agree with Michael Pollan's reflection in his essay "An Animal's Place" that future generations will look back on our treatment of these intelligent creatures in horror. ("Will history someday judge us as harshly as it judges the Germans who went about their ordinary lives in the shadow of Treblinka? Precisely that question was recently posed by J. M. Coetzee, the South African novelist, in a lecture delivered at Princeton; he answered it in the affirmative. If animal rightists are right, 'a crime of stupefying proportions' (in Coetzee's words) is going on all around us every day, just beneath our notice.")

Rowan Jacobsen's fascinating book about terroir is *American Terroir: Savoring the Flavors of Our Woods, Waters, and Fields* (Bloomsbury, 2010).

Soil

In *A Sand County Almanac*, Aldo Leopold wanted to show how nature sustains us. To do so, he chose the figure of a pyramid—and at the base he placed "soil." By setting soil as the foundation of his "biotic pyramid," Leopold attempted to demonstrate the simple fact that everything else—plants, insects, birds, mammals—depends upon soil for existence. "No ecologist before had presented such a comprehensive and comprehensible concept of the land and explained its implications for the broad range of conservation concerns," says biographer Julianne Warren. "This new understanding placed on people a new obligation to conserve soil and all that went with it."

Is soil the same as dirt? Clever headlines like "Are we treating our soil like dirt?" clearly make a distinction. But I have found that 1) most people use the words interchangeably and don't consciously think there's a difference, and 2) that soil scientists and conservationists use the word "soil." My own sense is that while soil clearly refers to the living substance on which life depends, dirt can but often doesn't. For example, when I dig my hands into a garden, that's soil; when I drive by a construction site, that's dirt. Another way to think of it is that soil is living dirt, and dirt is dead soil. In the end, I think "dirt" is useful as a broader term, because most people have an immediate image in mind. But soil is really what we're talking about—the living ground.

Find out more about the Soil Society of America (and of course their favorite subject, soil) at https://www.soils.org. To learn more about soil profiles, see the "Planting Seeds" blog from the California Department of Food and Agriculture at http://plantingseedsblog.cdfa .ca.gov/wordpress/?p=7930.

A reading list for anyone interested in learning more about soil would certainly include the following books: *Dirt: The Erosion of Civilizations,* by David R. Montgomery (University of California, 2007); *Dirt: The Ecstatic Skin of the Earth,* by William Bryant Logan (Riverhead, 1995); *Dirt: A Love Story,* edited by Barbara Richardson (ForeEdge, 2015); *Tales from the Underground: A Natural History of Subterranean Life,* by David W. Wolfe (Basic, 2001); *Life in the Soil,* by James B. Nardi (University of Chicago, 2007); *Cows Save the Planet: And other Improbable Ways of Restoring Soil to Heal the Earth,* by Judith D. Schwartz (Chelsea Green, 2013); and *The Soil Will Save Us: How Scientists, Farmers, and Foodies Are Healing the Soil to Save the Planet,* by Kristin Ohlson (Rodale, 2014).

The figure that the world has on average only six more decades of growing crops comes from the UN's Food and Agriculture Organization. To find out more, see "Only 60 Years of Farming Left If Soil Degradation Continues" from *Scientific American* at http://www.scientificamerican .com/article/only-60-years-of-farming-left-if-soil-degradation -continues/, and "We're treating soil like dirt" from *The Guardian*'s George Monbiot at https://www.theguardian.com/commentisfree/2015/mar/25/ treating-soil-like-dirt-fatal-mistake-human-life. To learn more about how "Europe's environmental laws overlook vital soil," see http://www

.dw.com/en/global-ideas-soil-erosion-agriculture-europe/
a-18807131.

For a sample of recent articles detailing good news about soil
and farming, see "A Sustainable Solution for the Corn Belt" by Mark
Bittman at http://www.nytimes.com/2014/11/19/opinion/a-sustainable
-solution-for-the-corn-belt.html; "Healthy Ground, Healthy Atmosphere: Recarbonizing the Earth's Soils" at http://ehp.niehs.nih.gov/
124-a30/; "Cover Crops, a Farming Revolution with Deep Roots in the
Past" at http://www.nytimes.com/2016/02/07/business/cover-crops-a
-farming-revolution-with-deep-roots-in-the-past.html; and "Agroecol-
ogy and industrial farming: leveling the playing field" at https://food
first.org/agroecology-and-industrial-farming-leveling-the
-playing-field/.

Ames

Joseph Wood Krutch is perhaps best known for his lovely book *The
Desert Year*, published in 1952.

The statistics about Iowa's transformation are stunning: "150 years
ago, nearly 85% of the state was tall grass prairie," explains the Iowa
Natural Heritage Foundation, "today less than 0.1% of this prairie
remains"; see http://www.inhf.org/ec1-prairie-management.cfm. The
extension service of Iowa State University reports that more than 37
percent of the state is planted in corn, and more than 25 percent in
(soy)beans. As of 2013 data, 91 percent of Iowa land consists of farms,
leaving only approximately 9 percent to interstates, roads, surface
water, incorporated land, or federal land; see http://www.extension
.iastate.edu/soils/crop-and-land-use-statewide-data.

I spent a fascinating afternoon at the Tallgrass Prairie Center with
its director, Laura Jackson. When I mentioned to her that some people
I'd told about Iowa's transformation couldn't believe the news, she
responded, "They need to drive through northern Iowa on a country
road." When she first did that herself, when she first came to northern
Iowa more than twenty years ago, it struck her as a wasteland. "Oh my
gosh, it's just really extreme," she said with a laugh. "There aren't many
places in the world that are like this. To be able to drive five hours east,
five hours west, five hours north, and two hours south before you hit
any expanse of natural lands at all—find another place on the map like
that. It doesn't exist."

For Jackson, this monotony is no mistake. In her article "Who
'designs' the agricultural landscape?" she argues that Iowa has been
systematically "designed primarily by global agribusiness corpora-

tions." Local farmers and communities have little power to resist such design. And the one group that could resist—urban consumers—are seduced by what she calls "the myth of the farmer as designer" and "remain ignorant of the aesthetic, ecological, and social consequences of their own appetites." Read more at http://lj.uwpress.org/content/27/1/23.refs.

To find out more about the Tallgrass Prairie Center and its efforts to preserve Iowa's prairie biodiversity, visit www.tallgrassprairiecenter.org.

My thanks to Mark Rasmussen of the Leopold Center for Sustainable Agriculture at Iowa State University. The challenge, he told me during my visit to his campus, is to get industrial agriculture to "quit fighting this defensive siege mentality, where they're just circling the wagons, trying to hold everyone off," rather than see that farming can't continue the course it's on.

For more on Fred Kirschenmann's writings, see https://www.leopold.iastate.edu/content/writings-fred-kirschenmann.

As we left Marsden Farms and began thinking about dinner, Fred Kirschenmann told me about his current activities at the Stone Barns Center in upstate New York. "Part of the mission there is to educate children. We have about ten thousand kids come through there a year, and the staff always brings back stories. My favorite is a school bus came out with some eight-year-old kids one morning. The first thing our guide did was took them to the garden so they could see where food comes from. Then at one point she reached down, pulled a carrot out of the ground, washed it off, and handed it to the kids, and said, 'Have a bite to see how great it tastes.' And she said this one little boy looked her right in the eyes and said, 'Oh, gross. Who stuck those in the dirt? Now we can't eat them.'"

Find out more about the Stone Barns Center at www.stonebarnscenter.org and its famous chef, Dan Barber, at www.bluehillfarm.com.

For a fascinating discussion of the "true cost" of a cheeseburger, see Mark Bittman's "The True Cost of a Burger" at http://www.nytimes.com/2014/07/16/opinion/the-true-cost-of-a-burger.html. The Story of Stuff Project has a good article on the concept of externalized costs at http://storyofstuff.org/blog/talk-about-externalized-costs/, an understanding of which is a requirement for anyone weighing the value of "the environment" compared to anything else.

There is so much happening with soil these days. On one hand, it seems most people know little about its value. On the other hand, the news is filled with stories about this value. Here are just a few recent headlines: "Human security at risk as depletion of soil accelerates, scientists warn";

"World loses trillions of dollars' worth of nature's benefits each year due to land degradation"; and "Earth has lost a third of arable land in past 40 years, scientists say." For an excellent article detailing the many threats to life on Earth from a loss of soil, see "What If the World's Soil Runs Out?" at http://world.time.com/2012/12/14/what-if-the-worlds-soil-runs-out/.

The most prominent research on the possible decline of nutrients in fruits and vegetables has been led by Donald R. Davis of the University of Texas. Find out more in "Declining Fruit and Vegetable Nutrient Composition: What is the Evidence?" http://hortsci.ashspublications.org/content/44/1/15.full.

Find Wendell Berry's lecture in which he argues that "it all turns on affection" at http://www.neh.gov/about/awards/jefferson-lecture/wendell-e-berry-lecture. He writes, "The word 'affection' and the terms of value that cluster around it—love, care, sympathy, mercy, forbearance, respect, reverence—have histories and meanings that raise the issue of worth. We should, as our culture has warned us over and over again, give our affection to things that are true, just, and beautiful. When we give affection to things that are destructive, we are wrong. A large machine in a large, toxic, eroded cornfield is not, properly speaking, an object or a sign of affection."

Fireflies are increasingly endangered around the world, and nowhere more than here in the United States. Find out more at http://www.firefly.org/how-you-can-help.html.

Grass

For more on Walt Whitman's "A child said, What is the grass?" see https://www.poets.org/poetsorg/poem/child-said-what-grass.

Find Michael Pollan's essay "Why Mow?" at http://michaelpollan.com/articles-archive/why-mow-the-case-against-lawns/. "Nowhere in the world are lawns as prized as in America," he writes. And that "Lawns appear to cover more than three times the number of acres that irrigated corn covers" reports a NASA article, "Lawn Surface Area in the United States" at http://earthobservatory.nasa.gov/Features/Lawn/lawn2.php. Perhaps the most important study on American lawns to date was conducted in 2003 by NASA's Cristina Milesi, who used satellite data to determine how much lawn covered the country. A new study in 2014 focused on six American cities to determine Americans' lawn-care habits and found that "79% of surveyed residents watered their lawns and 64% applied fertilizer"; see https://www.soils.org/discover-soils/story/national-study-reveals-urban-lawn-care-habits. "These numbers are important," wrote Peter Groffman, one of the study's authors, because "what we do in our suburban and urban yards has a big impact, for better or worse, on the environment."

Find out more about Shay and Eric Lunseth, and organic lawn care in general, at http://www.organiclawnsbylunseth.com.

Details of how a CDC study found "pesticides in 100% of the people who had both blood and urine tested" can be found at http://learn .eartheasy.com/2009/01/lawn-care-chemicals-how-toxic-are-they/.

To find out more about Roundup Ready Kentucky bluegrass, see "Roundup Ready Kentucky Bluegrass: Benefits and Risks," at http://turf .umn.edu/2014/04/roundup-ready-kentucky-bluegrass-benefits-risks/. Glyphosate, the main chemical used in Roundup, remains highly controversial. For a sample of the debate, see such articles as "World Health Organization's new Q&A on Glyphosate Confirms Toxicity of Round Up" at http://www.globalresearch.ca/world-health-organizations-new -q-a-on-glyphosate-confirms-toxicity-of-round-up/5513497; "Experts call on feds to re-evaluate the world's most heavily used herbicide" at http://www.environmentalhealthnews.org/ehs/news/2016/feb/ glyphosate-roundup-monsanto-cancer-endocrine-disruptor-science; and "Misgivings about how a weed killer affects the soil" at http://www .nytimes.com/2013/09/20/business/misgivings-about-how-a-weed -killer-affects-the-soil.html.

Perhaps the best-known negative effect of Roundup and glyphosate is the devastation scientists say it has caused to the world's monarch butterfly population. For more, see "Limits sought on weed killer glyphosate to help monarch butterflies" at http://articles.latimes.com/2014/feb/25/science/ la-sci-sn-monarch-butterfly-roundup-20140224, and https://www.nrdc .org/experts/sylvia-fallon/monarch-butterflies-plummet-its-time-rethink -widespread-use-our-nations-top. But around the world, insect populations in general have seen an enormous decline in their numbers, mostly due to habitat loss and climate disruption. A wonderful source for learning more is "Vanishing Act: Why Insects Are Declining and Why It Matters," from Yale Environment 360. The average of a 45 percent decline in insect populations over the past thirty-five years boggles the mind.

When I asked Janet Davis of Charlottesville, Virginia's Hill House Farm and Nursery about the value of lawns for pollinating insects, she laughed. "Virtually none," she said. "What we need is grass that has space between the clumps, and we make it so thick it's like a putting green. A lot of our insects and a lot of our important pollinators are ground nesting, and they need a little bit of bare dirt. If they can't get to that, they're gone. Most of them only fly a hundred yards or less, and they only live a short period of time. So they can't have big expanses of lawn. There's nothing to eat. They're rapidly disappearing because of our obsession."

This summer, in the back yard of the house I'm renting in Minneapolis, there is clearly a bumblebee nest between clumps of turfgrass. I watch the big bees coming in to land, maneuvering like tiny yellow-black cargo helicopters swaying in the breeze as they settle down between the green blades of grass. I am quite certain that I have never before noticed bumblebees nesting on the ground. And I am amazed.

Type "environmental costs of golf courses" into any web browser and no shortage of articles will arrive. Among them, "America's 18,000 Golf Courses Are Devastating the Environment," which though somewhat dated (2004) still wins the day for most alarming headline. A more recent article, "Health of ecosystems on U.S. golf courses better than predicted, researchers find," at https://www.sciencedaily.com/releases/2014/04/140410122201.htm, reflects a growing consensus that perhaps golf courses are not quite so bad as is commonly thought.

The Wes Jackson book I quote is *Becoming Native to This Place* (University Press of Kentucky, 2004). But he is also the author of several other important books, as well as the founder of The Land Institute in Salina, Kansas. The Land Institute works "to ensure food security by replacing the current extractive and chemically intensive model for agriculture with a sustainable model inspired by nature." Find out more at landinstitute.org.

Sandhills

Peter Matthiessen is the author of several wonderful books about the natural world (as well as several wonderful novels). They include *The Birds of Heaven* (from which this chapter's epigraph comes), *The Tree Where Man Was Born* (about East Africa's Serengeti), and *The Snow Leopard*, which won the National Book Award in 1979.

The study detailing the deteriorating state of Minnesota's wetlands can be found at https://www.pca.state.mn.us/research-shows-minnesota-wetlands-healthy-overall-suffering-some-regions-2015-october. In general, despite efforts to protect them, wetlands continue to decline in quality and number. The EPA reports, "Despite all the benefits provided by wetlands, the United States loses about 60,000 acres of wetlands each year." Most of the laws to protect wetlands have come since 1984, when Congress realized that more than half of US wetlands had already been drained or filled for development.

All the quotes reflecting the beauty of the tallgrass prairie can be found in *The Tallgrass Prairie Reader*, edited by John T. Price (University of Iowa, 2014). This is a beautiful book full of firsthand accounts of what was once a magnificent ecosystem.

The threats to the remaining native grasslands continue, and in recent years have increased. "The Enormous Threat to America's Last Grassland" reads the *Washington Post* headline (June 16, 2016). The subtitle to another article reads, "Across the northern plains, native grassland is being turned into farmland at a rate not seen since the 1920s. The environmental consequences could be disastrous"; see http://prospect.org/article/plowed-under. These are the grasslands upon which ducks, cranes, and other birds directly rely. Incredibly, reports the Prospect.org article, "the rates of land-use change in the region...parallel the deforestations taking place in Brazil, Malaysia, and Indonesia."

For an excellent article on the sandhill crane migration passing through Nebraska's Platte River, see Alex Shoumatoff's "500,000 Cranes Are Headed for Nebraska in One of Earth's Greatest Migrations," from *Smithsonian* magazine. To find out more about the Nicholson Audubon Center at Rowe Sanctuary, visit http://rowe.audubon.org. Even better, go in person between March and early April to see for yourself one of nature's most breathtaking scenes.

Another excellent article, though on "Great Migrations" in general, appeared in *National Geographic*: "Great Migrations: Birds, butterflies, and beasts take off. Humans interfere," by David Quammen, November 2010.

The quote from Linda Hogan comes from her book *Dwellings: A Spiritual History of the Living World* (Norton, 1995).

About the ability of sandhill cranes to survive serious weather, Gary Krapu told me, "They've evolved so that they can migrate long distances with a minimum of energy being expended riding wind uplifts, and they don't make many mistakes. Sometimes when crossing the Bering Strait into Russia, if there's fog and cold rain, they can lose their orientation and go into the waves, and you'll get drifts of dead cranes coming onto the shoreline."

But the place that concerns Krapu most is the Platte River, "because they are there in huge numbers and it's at a time of the year that you can get real severe ice storms and blizzards. I've been out in blizzards where you can hardly see, and they'll have their head under their wing, and you wouldn't hardly know there was a bird out there—you'll see bumps on the landscape." The cranes can usually survive that type of weather, he says, but if snowstorms turn into ice storms, and the ice starts sticking to their feathers, that's another story. "That doesn't occur very often, but when you have a situation like the Platte where you have 80 percent of the population in one spot, a half million cranes, nature generally did not plan for that."

For more about cranes, visit the International Crane Foundation in Baraboo, Wisconsin, or online at www.savingcranes.org.

Find out more about impacts to the US National Wildlife Refuge System in "Housing development will limit the ability of the National Wildlife Refuge System to respond to climate change at http://conserva tioncorridor.org/2016/07/housing-development-will-limit-the-ability -of-the-national-wildlife-refuge-system-to-respond-to-climate-change/. For similar information about development around national parks, see http://www.nationalparktraveler.com/2012/04/study-development -around-national-parks-far-surpasses-other-parts-country9678. And for an excellent article on similar situations around the world, see "How Nations are Chipping Away at Their Protected Lands" at http://e360 .yale.edu/feature/how_nations_are_chipping_away_their_protected _lands/2989/.

Aldo Leopold's "Thinking Like a Mountain" appears in *A Sand County Almanac* (Oxford, 1948).

Appalachia

An earlier version of this chapter appeared in *Fracture: Essays, Poems, and Stories on Fracking in America*, edited by Taylor Brorby and Stefanie Brook Trout (Ice Cube, 2016).

Rachel Carson's *The Sea Around Us* was originally published in 1951. A national best-seller, it won the John Burroughs Award for Nature Writing in 1952. Its success and the success of her other books such as *Under the Sea-Wind* (1941) and *The Edge of the Sea* (1955) created the space for the book for which she is today best known, *Silent Spring* (1962).

Curiously enough, "Hydrofracking 101" has now disappeared from the Halliburton website.

For more information on the destruction caused to ground by frack-ing, see "Thirty Thousand Square Kilometers of Land Lost to Oil and Gas Development" at http://www.sciencemag.org/news/2015/04/thirty -thousand-square-kilometers-land-lost-oil-and-gas-development, and "New research on cumulative ecological impact of oil and gas" at https:// www.hcn.org/articles/researchers-calculate-cumulative-ecological -impact-of-oil-and-gas-boom-and-its-big).

You don't need to know how to spell "Hieronymus" to search for the Dutch painter of the fifteenth and sixteenth centuries—just begin typing "H-i-e-r-o" and your search engine will probably know what you mean. Housed at the Prado Museum in Madrid, Spain, *The Garden of Earthly Delights* is perhaps his most famous work. To learn more (and

see images of his works), see http://www.hieronymus-bosch.org/the
-complete-works.html.

William Styron shares his harrowing personal story of clinical
depression in *Darkness Visible: A Memoir of Madness*, published in 1992.
The book is, perhaps counterintuitively, a beautifully written account
of a descent into hell.

Cormac McCarthy's *The Road* won the Pulitzer Prize for fiction in
2007. Not long after its publication, the British journalist George
Monbiot wrote, "A few weeks ago I read what I believe is the most
important environmental book ever written. It is not *Silent Spring*,
Small is Beautiful or even *Walden*." He was talking about *The Road*.

For an excellent summary of how pumping toxic liquid from drilling operations back into the ground poses possible risks to drinking
water, see Abrahm Lustgarten, "Injection Wells: The Poison Beneath
Us" at https://www.propublica.org/article/injection-wells-the-poison
-beneath-us.

Groundwater is, yes, water in the ground. But more specifically it's
water found in the tiny spaces between soil and rock. Larger openings
underground—think caves and caverns, or the remains of lakes under
Mexican capitals—that are filled with groundwater are known as aquifers. In most cases, water drawn from underground aquifers and
sprayed onto crops or lawns, or drawn from drinking taps, or allowed
to run down streets on hot summer days, is water that has been underground for centuries. The fact that this resource lies hidden below us
creates a challenge. For example, while research does agree that the
Ogallala Aquifer has been lowered significantly by decades of use, it
does not agree on how much the aquifer still holds or when it will run
dry. Because so much of Central Plains agriculture depends on the
Ogallala for its water, how much is left matters a great deal. And that's
only one major aquifer—around the world other major aquifers have
also been drawn down to the point where experts now fear they will
soon be exhausted.

Scientists using NASA satellites have found that fully twenty-one
of Earth's thirty-five largest aquifers had "passed their sustainability
tipping points, meaning more was removed than replaced during the
decade-long study period." The study was the first to show definitively
how the globe's aquifers are struggling to keep up with the demands
from human activities such as mining and agriculture, and Earth's
ever-growing human population. Jay Famiglietti of NASA's Jet Propulsion Laboratory in California put it succinctly: "The situation is
quite critical. The water table is dropping all over the world. There's
not an infinite supply of water."

Here in the United States, in addition to that of the Ogallala Aquifer, the most disconcerting groundwater losses are taking place as a result of increasing human populations in places such as Nevada, Utah, and Colorado, and in the agriculture-intensive Central Valley in California. For decades, energy and mining companies have used aquifers as dumping grounds, polluting what are often pristine water sources beyond the capacity for renewal. The oil and gas industry is by far the biggest user of underground aquifers and by far the biggest polluter. But, said a senior EPA official, "right now nobody — nobody — wants to interfere with the development of oil and gas or uranium. The political pressure is huge not to slow that down."

Perhaps best known for her book *Living Downstream* (1997), the story of linking her bladder cancer with the industrial chemicals flowing from a factory upstream, Sandra Steingraber has now turned her considerable talents toward the issue of fracking. "I have come to believe that extracting natural gas from shale using the newish technique called hydrofracking is *the* environmental issue of our time," she wrote. "And I think you should, too." Read more at "The Whole Fracking Enchilada" at https://orionmagazine.org/article/the-whole-fracking -enchilada/. I often tell students and friends that Steingraber's *Having Faith* (2001) — her story of becoming pregnant for the first time at age thirty-eight while studying the effects of toxins in the environment — was the most powerful book I read while in graduate school.

The Collapse of Western Civilization: A View from the Future (2014) is Naomi Oreskes's second book with Erik M. Conway. Their first book together, *Merchants of Doubt: How a Handful of Scientists Obscured the Truth on Issues from Tobacco Smoke to Global Warming*, was published in 2010. A talented speaker and writer on the subject of climate change, Oreskes is a professor of the History of Science at Harvard University. For more information, including links to her writings and interviews, visit http://histsci.fas.harvard.edu/people/naomi-oreskes.

John Clare lived from 1793 to 1864, spending nearly his entire life close to his childhood home of Helpston, England. Geoffrey Summerfield's enlightening introduction to Clare's poetry appears in *John Clare: Selected Poems* (Penguin, 2004). For an excellent discussion of Clare's poetic significance, see Gary Harrison's essay "John Clare's Poetics of Acknowledgement" at http://www.euppublishing.com/doi/ abs/10.3366/rom.2012.0063?journalCode=rom. For Harrison, Clare's poetry dissolves the dualism that so commonly underlies the Western approach to nature. Clare's "astonishing love for Helpston and its environs," his "habits of close observation and precise detail," and use of "sensuous immediacy" to describe the natural world all add up to "his propensity to act and speak in its behalf." Harrison describes "our refusal to acknowledge the bond we hold with the earth," and argues that "it is not that we lack knowledge about environmental ills, but that we fail to acknowledge and act upon that knowledge." In other words, "We fail to acknowledge nature and our responsibility to... see it, acknowledge it in all its otherness and mystery, and try to help others to do the same."

Treblinka

The chapter's epigraph comes from *The Forest Unseen: A Year's Watch in Nature*, by David George Haskell (Penguin, 2012).

Before reading Jean-François Steiner's *Treblinka*, about the prisoner uprising that finally brought an end to the Nazi death camp, I knew of the camp only generally. By the time I had finished Steiner's story, I knew I had to see this place for myself. Reading Vasily Grossman's *The Hell of Treblinka*, originally published in 1944, further solidified my desire. My first goal was simply to see what it felt like to walk the grounds where something so horrible had taken place.

My deep thanks to Shiri Sandler of the Museum of Jewish Heritage in New York. Even before the extermination camps were created at Treblinka, Bełżec, and Sobibór, she told me, the Einsatzgruppen (German for "task forces" or "deployment groups"), mobile killing squads, killed about 1.5 million people. "And so in the areas around these camps," she described, "you already have people being gathered together and taken to fields outside their village and being shot." Of Treblinka she said, "there were actually two camps at Treblinka, a work camp where many died, and an extermination camp where everyone died."

"My second name is Beth," she told me, "in honor of my German great-grandfather who was a German Jewish veteran of World War One, a decorated veteran. And my whole childhood my middle name was used, I was Shiri-Beth, for Benjamin. We have a picture of him in his uniform, with his medals. He was killed on Christmas 1944, shot in a subcamp of Dachau. Survivors of the camp, I've talked to one of them, have said a group of men were shot as a Christmas present to Hitler."

Sandler tells me of a photograph at Birkenau that stays with her. It's of a woman in one of the Hungarian deportation pictures that you see on one of the plaques there. "She was probably beautiful, she has dark hair and is wearing it down, and she's wearing what looks like a cape. And she looks ravaged—she looks thin and she's drawn and she's haunting. And she's just waiting," Sandler explains. "In those pictures they're just sitting in the woods waiting for their turn in the gas chambers."

I think of Sandler telling me about her anger at the language used at some German memorials, "trying to retroactively give agency or humanity." Of Berlin's Memorial to the Murdered Jews of Europe she told me, "I was there with my husband, and he was like, 'Is this clear enough for you, the language? Or do you need it to be'—and this is his black humor—'the Memorial to the Murdered, Gassed, Beaten, Starved, Raped, and otherwise Destroyed Jews of Europe?' And I was like, 'Well, that would be better, but unwieldy. This is okay. 'Murdered' is so clear.'"

When I travel for research, I carry my digital recorder wherever I go, mostly to record other voices but sometimes to record my thoughts, knowing that I will listen later, when writing. But the recordings I make at Treblinka are a few questions at first to Josef, and then only the sound of me, alone, walking on this ground, at this site, with long spaces of no words, of—I don't want to say silence, because I can hear the wind and the scattered songbirds, and my feet on the gravel and grass. But no words. And finally I turned the recorder off, because I had all the wind and pine trees and birdsong I could use, and it felt impossible to speak.

At Treblinka, at the edge of the forest clearing, a series of black-and-white photographs have been blown up to poster size, the first of an enormous steam shovel on tank tracks with the gas chambers in the background. Beneath the photo the caption reads, "The presented pictures were taken by Kurt Franz, the deputy commandant of the death camp in Treblinka. They come from the album called 'Beautiful Times.'"

The one individual name on a jagged stone marker at Treblinka belongs to Janusz Korczak. For many years he worked as the director of a Jewish orphanage in Warsaw. The story of his refusal of several opportunities to abandon his orphans and escape their fate—and of his leading the column of nearly two hundred children dressed in their finest clothes to the cattle cars—is heartbreaking. He died with his orphans at Treblinka in August 1942.

For an excellent documentary on Caroline Sturdy Colls and her work at Treblinka, see "Treblinka: Hitler's Killing Machine" by the Smithsonian Channel at http://www.smithsonianchannel.com/shows/treblinka-hitlers-killing-machine/0/3403868.

Along with Ivar Schute, two other archeologists were involved in the work at Sobibór: Wojciech Mazurek (Poland) and Yoram Haimi (Israel). When I asked Schute how he and his fellow archeologists at

Sobibór could be sure they had found the gas chamber remains, he told me, "The spatial structure of the site makes it clear. From the ramp a wooden fenced corridor of about two hundred meters, which we found, leads to a camp-within-a-camp, which consists mainly of mass graves, which we found. The connection between the corridor known, as at Treblinka, as the *Himmelfahrt Strasse* should be the gas chamber complex. And that is exactly where we found the brick building."

In talking with Schute, I was reminded of the incredible toll inflicted on Dutch Jews during the Holocaust. If ever we in the West need a reminder that, as Shiri Sandler told me, "it was exactly what it would feel like for us," looking at the photos of these Dutch citizens dressed in their best clothes boarding cattle cars ought to do the trick. For me, the people being sent to the extermination camps look exactly like my grandparents in 1940s southern Minnesota.

While the story of Anne Frank is known around the world, many fewer of us know that in total, more than 56,500 Dutch Jews were deported to Auschwitz, where only about 1,000 survived. More than 34,000 Dutch Jews were sent to Sobibór, where fewer than 20 survived. In total, of the more than 107,000 Dutch Jews deported, at least 102,000 perished. When you're in Amsterdam, the National Holocaust Museum and the Jewish Historical Museum are not to be missed.

Alaska

A few more facts about Alaska: the largest state in the Union, it is larger than Texas, California, and Montana (numbers 2, 3, and 4) combined; Alaska has more than 50 percent of the US coastline; and only 20 percent of the state's roads are paved, compared to an average of 91 percent in the other forty-nine states.

Find out about the Yukon-Kuskokwim Delta at www.fws.gov/ref uge/yukon_delta. Or, better yet, go see it for yourself: the flight from Anchorage to Bethel takes between 75 and 105 minutes, and Alaska Airlines has six flights daily. Driving there isn't an option.

For an excellent article on the future of the Yup'ik people, see "Baked Alaska: A Snowless Climate Threatens the Survival of the Yupik People of Togiak," at http://www.newsweek.com/2015/06/12/ togiak-alaska-lacks-snow-338600.html.

For more on Woods Hole Research Center and its work on climate change, especially thawing permafrost, visit www.whrc.org. Listen to Dr. Susan Natali discuss thawing Alaskan permafrost on the radio program *Living on Earth* at http://www.loe.org/shows/segments.html? programID=15-P13-00024&segmentID=1. And if you're interested in going straight to the scientific report, see "Climate change and the permafrost carbon feedback," in *Nature* 520 (April 9, 2015): 171–179. For the Woods Hole basic primer on the issue, see http://whrc.org/wp-content/uploads/2015/06/PB_Permafrost.pdf). Among the implications listed: "Carbon emissions from thawing permafrost accelerate climate warming, so the potential exists for a catastrophic, self-reinforcing cycle of warming and thawing permafrost."

I thank Norman Wirzba for a wonderful long conversation about soil and the sacred. Find out more about him at https://divinity.duke.edu/faculty/norman-wirzba. "The biblical story is telling us that the fate of soil and humanity are inextricably intertwined," he writes in "Dramas of Love and Dirt: Soil and the Salvation of the World." When soil suffers, he says, "so do we."

The Sierra Nevada

I first read the phrase "intimate to the degree of being sacred" in Sandra Steingraber's *Having Faith: An Ecologist's Journey to Motherhood* (2001). She was talking about breastfeeding.

Read more from Joanna Eede in her *National Geographic* blog http://voices.nationalgeographic.com/2011/04/01/uncontacted-tribes-the-last-free-people-on-earth/. The photographs are amazing too.

At only fifty-eight pages, Barry Lopez's brilliant *The Rediscovery of North America* (Vintage, 1992) is a quick and powerful read. Part of the power comes from his subject, but it is Lopez's clear and thoughtful language that makes his argument resonate.

Learn more about Kathleen Dean Moore at www.riverwalking.com. I first met Kathy while at school in Nevada. She had come over from Corvallis to visit our Literature and Environment program at the university. I was immediately taken by her thoughtful, serious, joyous approach to our world's environmental problems. When I first had the idea for my anthology on light pollution, the University of Nevada Press said, Find a few writers willing to contribute and then we'll talk contract. I immediately thought of Kathleen Dean Moore, and when she said yes to my invitation and sent me the beautiful "The Gifts of Darkness," I knew I would soon have my anthology. I am lucky to call her a friend, and I hold her in great admiration for the work she is doing to raise awareness about the urgent need to address climate change.

You can find out more about Dr. Miles Silman at http://college.wfu.edu/biology/people/faculty/silman/, and in Elizabeth Kolbert's best-selling book *The Sixth Extinction*, in which he's featured.

For a good discussion of the most recent UN projections on human population growth, see "Global population set to hit 9.7 billion people by 2050 despite fall in fertility," in *The Guardian* (July 29, 2015).

For more information on Wayne Roosa, including images of his artwork, visit www.wayneroosa.com. Learn more about the exhibition at the Minneapolis Institute of Art titled *Sacred* by visiting http://new.artsmia.org/sacred/. And find out more about the artist Francis Alÿs at www.francisalys.com.

And "then the Lord God formed man from the dust of the ground," reads Genesis 2:7. And "out of the ground the Lord God made to grow every tree that is pleasant to the sight and good for food," Genesis 2:9. That "dust to dust" idea is one we repeat, but here it's really from ground to ground, as in, "until you return to the ground, for out of it you were taken," Genesis 3:19. I say this because, of all that has become clear to me about our relationship with the ground, this is no doubt true: ground is home, and heaven too.

Luna showed me this one evening in the North Carolina woods. We had moved to Winston-Salem and I'd found us a county park to walk and run in, the two of us alone. And on this day we were running, the autumn humidity covering my shirtless body in sweat. We began the last descending trail winding through the woods, Luna about ten feet in front of me, leading the way, her orange-and-white ears flaring now and then as she galloped along the trail. We both were still young enough and strong enough to be running through the woods, and now on the downward slope toward the tail end of the run, I thought, *This is heaven. This. Here. Now.*

Index

About the Author

PAUL BOGARD is the author of *The End of Night* and editor of the anthology *Let There Be Night: Testimony on Behalf of the Dark*. His writing and commentary on the natural world has appeared in the *Los Angeles Times*, and on *Slate*, *Salon*, and *All Things Considered*. He teaches creative nonfiction at James Madison University and lives in Virginia and Minnesota.